AF839704

Comparing Cell Culture and Mouse Assays for Measuring Infectivity of *Cryptosporidium*

Awwa Research Foundation

About the Awwa Research Foundation

The Awwa Research Foundation (AwwaRF) is a member-supported, international, nonprofit organization that sponsors research to enable water utilities, public health agencies, and other professionals to provide safe and affordable drinking water to consumers.

The Foundation's mission is to advance the science of water to improve the quality of life. To achieve this mission, the Foundation sponsors studies on all aspects of drinking water, including supply and resources, treatment, monitoring and analysis, distribution, management, and health effects. Funding for research is provided primarily by subscription payments from approximately 1,000 utilities, consulting firms, and manufacturers in North America and abroad. Additional funding comes from collaborative partnerships with other national and international organizations, allowing for resources to be leveraged, expertise to be shared, and broad-based knowledge to be developed and disseminated. Government funding serves as a third source of research dollars.

From its headquarters in Denver, Colorado, the Foundation's staff directs and supports the efforts of more than 800 volunteers who serve on the board of trustees and various committees. These volunteers represent many facets of the water industry, and contribute their expertise to select and monitor research studies that benefit the entire drinking water community.

The results of research are disseminated through a number of channels, including reports, the Web site, conferences, and periodicals.

For subscribers, the Foundation serves as a cooperative program in which water suppliers unite to pool their resources. By applying Foundation research findings, these water suppliers can save substantial costs and stay on the leading edge of drinking water science and technology. Since its inception, AwwaRF has supplied the water community with more than $300 million in applied research.

More information about the Foundation and how to become a subscriber is available on the Web at **www.awwarf.org**.

Comparing Cell Culture and Mouse Assays for Measuring Infectivity of *Cryptosporidium*

Prepared by:
Paul A. Rochelle
Metropolitan Water District of Southern California, La Verne, CA 91750

Marilyn M. Marshall
University of Arizona, Tucson, AZ 85721

Jan Mead
Atlanta Veterans Medical Center, Atlanta, GA 30033

and

Jeffrey S. Rosen
Perot Systems Government Services, Scituate, MA 02066

Jointly sponsored by:
Awwa Research Foundation
6666 West Quincy Avenue, Denver, CO 80235-3098

and

U.S. Environmental Protection Agency
Washington D.C.

Published by:

Awwa Research Foundation
Advancing the Science of Water®

IWA Publishing

DISCLAIMER

This study was funded by the Awwa Research Foundation (AwwaRF). AwwaRF assumes no responsibility for the content of the research study reported in this publication or for the opinions or statements of fact expressed in the report. The mention of trade names for commercial products does not represent or imply the approval or endorsement of AwwaRF. This report is presented solely for informational purposes.

Published by IWA Publishing, Alliance House, 12 Caxton Street, London SW1H 0QS, UK

Telephone: +44 (0) 20 7654 5500; Fax: +44 (0) 20 7654 5555; Email: publications@iwap.co.uk;
Web: **www.iwapublishing.com**

AwwaRF report number 91027F. Originally published by AwwaRF for its subscribers in 2004.
IWA Publishing version published 2005.

ISBN 1 84339 915 6

CONTENTS

TABLES

FIGURES

FOREWORD

The Awwa Research Foundation is a nonprofit corporation that is dedicated to the implementation of a research effort to help utilities respond to regulatory requirements and traditional high-priority concerns of the industry. The research agenda is developed through a process of consultation with subscribers and drinking water professionals. Under the umbrella of a Strategic Research Plan, the Research Advisory Council prioritizes the suggested projects based upon current and future needs, applicability, and past work; the recommendations are forwarded to the Board of Trustees for final selection. The foundation also sponsors research projects through the unsolicited proposal process; the Collaborative Research, Research Applications, and Tailored Collaboration programs; and various joint research efforts with organizations such as the U.S. Environmental Protection Agency, the U.S. Bureau of Reclamation, and the Association of California Water Agencies.

This publication is a result of one of these sponsored studies, and it is hoped that its findings will be applied in communities throughout the world. The following report serves not only as a means of communicating the results of the water industry's centralized research program but also as a tool to enlist the further support of nonmember utilities and individuals.

Projects are managed closely from their inception to the final report by the foundation's staff and large cadre of volunteers who willingly contribute their time and expertise. The foundation serves a planning and management function and awards contracts to other institutions such as water utilities, universities, and engineering firms. The funding for this research effort comers primarily from the Subscription Program, through which water utilities subscribe to the research program and make an annual payment proportionate to the volume of water they deliver and consultants and manufacturers subscribe based on their annual billings. The program offers a cost-effective and fair method for funding research in the public interest.

A broad spectrum of water supply issues is addressed by the foundation's research agenda: resources, treatment and operations, distribution and storage, water quality and analysis, toxicology, economics, and management. The ultimate purpose of the coordinated effort is to assist water suppliers to provide the highest possible quality of water economically and reliably. The true benefits are realized when the results are implemented at the utility level. The foundation's trustees are pleased to offer this publication as a contribution toward that end.

Walter J. Bishop
Chair, Board of Trustees
Awwa Research Foundation

James F. Manwaring, P.E.
Executive Director
Awwa Research Foundation

ACKNOWLEDGMENTS

The majority of the experimental work completed at the Metropolitan Water District of Southern California (MWD) was performed by Anne M. Johnson and Katrin Hanley. The authors thank other staff at the various laboratories for a variety of tasks: Mark Williams and Connie Chou for exposing oocysts to ozone; Alex Mofidi for assistance with UV exposures; Anagha Chitre, Daffodil Fallar, Patty Huang, Leigh Jacobs, Walter Naro, Julie Stewart, and Emily Sullivan for technical assistance in the laboratories; Dick G. Korich for help with animal studies and data analysis; Jose Sobrinho for help with the statistical analyses; Ricardo De Leon for technical advice; and Lynn Kelemen for project and contract administration.

We are also grateful to Cynthia Chappell (University of Texas Health Science Center, Houston, Texas), Michael Arrowood (Centers for Disease Control and Prevention, Atlanta, Georgia), Hugh Smith (Scottish Parasite Diagnostic Laboratory, Glasgow, Scotland), Steve Wright (Moredun Research Institute, Peniculk, Scotland), and Robert Gilman (Johns Hopkins University, Baltimore, Maryland) for providing isolates of *Cryptosporidium* for this study. We also thank David Polchert and Rebecca Hoffman (Wisconsin State Laboratory of Hygiene) for flow cytometry services.

The advice of the Project Advisory Committee (PAC)— Zia Bukhari (American Water Works Service Company), Nicholas Dugan (United States Environmental Protection Agency), Michael Hotaling (Harwoods Mill Water Treatment Plant), Norman Pieniazek (Centers for Disease Control and Prevention), Frank W. Schaefer, III (United States Environmental Protection Agency), and Keith Woods (Kansas State University) —and the help of the AwwaRF Project Officer, Misha Hasan, are also appreciated.

EXECUTIVE SUMMARY

The protozoan parasite *Cryptosporidium parvum* is common in many animal species and oocysts are frequently detected in untreated surface waters. Oocysts are resistant to chlorine disinfection at the concentrations typically applied during drinking water treatment and there are limited chemotherapeutic agents currently available for treatment of infected individuals. Although the disease is usually self-limiting in otherwise healthy patients, it can be fatal for those with weakened immune systems. Consequently, *C. parvum* presents a serious threat to public health and is a significant concern for the water industry. Over 55 documented outbreaks of cryptosporidiosis have been associated with either drinking water or recreational use of water since 1983. The largest waterborne outbreak to date occurred in 1993 in Milwaukee with estimates of the affected population ranging from 15,000 to 400,000 individuals and up to 100 deaths attributed to the outbreak. Various studies have demonstrated that over half of the source water in North America may contain *Cryptosporidium* oocysts.

Therefore, the water industry requires infectivity assays to provide much needed information on the occurrence of infectious *Cryptosporidium* oocysts in raw and finished water and the efficacy of disinfection and treatment strategies. The three techniques available to determine *C. parvum* infectivity are human volunteers, animal models, and in-vitro cell culture. Human volunteer studies are not practical for use on a routine basis by the water industry. Mouse infectivity assays are currently the accepted technique for measuring *C. parvum* infectivity (traditionally considered as the "gold standard") but the expense and labor intensive nature of mouse infectivity assays limits their practicality for routine monitoring. Also, some strains/isolates of *C. parvum* do not efficiently infect mice or other animal models. Consequently, in-vitro cell culture provides a promising and practical alternative technique for determining infectivity of waterborne *Cryptosporidium*. Cell culture techniques are routinely used for measuring the infectivity of a wide range of human pathogens and a variety of different cell culture systems and detection methodologies have been applied to *C. parvum*. However, the water industry does not yet recognize cell culture as an acceptable technique because the relationship between cell culture-based infectivity and animal infectivity has not been thoroughly investigated. Although many cell culture methods have been developed for *C. parvum*, it has not been proven that infectivity in cell culture is a good indicator of the ability of oocysts to cause infections in animals. Also, the apparent efficacy of disinfectants may be affected by the method used to measure inactivation; cell culture-based procedures may not give the same results as animal infectivity assays. In addition, different cell lines used for in-vitro cell culture demonstrate varying susceptibility to *C. parvum* infection. Therefore, studies are needed to assess the correlation between cell culture and animal-based infectivity of *C. parvum* to determine whether the different methods vary in their sensitivity to infection, dose responses, or ability to demonstrate oocyst inactivation.

A key concern in producing the amount of *Cryptosporidium* disinfection data which is needed for regulatory development or for testing disinfectant efficacy is the lack of a secondary reference method for *Cryptosporidium* inactivation. A study designed to determine which of four in-vitro viability surrogates could serve to replace animal infectivity demonstrated that none of the in-vitro surrogates were reliable indicators of infectivity (Bukhari et al., 2000). The results of this study indicated that for ozone and UV light, animal infectivity was the only reliable indicator for oocyst infectivity. This severely limits the amount of disinfection work, which can be done, as animal infectivity studies require specialized facilities found in only a limited number of laboratories and are extremely labor intensive and expensive. Consequently, alternative infectivity methods, such as cell culture, need to be adopted, but only if they demonstrate equivalency or superiority to animal models.

RESEARCH OBJECTIVES

Cell culture-based infectivity assays provide a useful tool that can be used to gain valuable information about *Cryptosporidium* in drinking water. They are generally practical and relatively straightforward techniques that can be used to measure the efficacy of disinfectants, the infectivity of oocysts detected in water, and the survival and persistence of oocysts in the environment. However, further research is necessary to compare cell culture with the "gold standard" of mouse infectivity before the method can be widely adopted by the water industry. Therefore, the overall objective of the research described in this report was to compare in-vitro cell culture-based systems with an animal model for determining the infectivity of *Cryptosporidium parvum* oocysts and to compare the assays for measuring disinfection of oocysts. The experimental aspects of the project were defined by the following specific objectives:

- Use at least four different isolates of *C. parvum* to compare the dose response and sensitivity of cell culture and mouse infectivity assays.
- Compare cell culture and animal infectivity assays for determining the efficacy of chemical (ozone) and physical (UV) disinfectants.
- Perform an experimental assessment of ozone inactivation of *C. parvum*.
- Compare infectivity methods with oocysts exposed to environmental water samples.
- Determine the reproducibility and variability of the methods by statistical analysis of all data.

APPROACH

The infectivity of five isolates of *C. parvum* and one isolate of *C. hominis* (previoulsy referred to as genotype 1 *C. parvum*; Morgan-Ryan et al., 2002) was measured in CD-1 mice and three cell culture-based assays: HCT-8 cells with infection detected using a *C. parvum*-specific reverse transcriptase-polymerase chain reaction (RT-PCR); Caco-2 cells with RT-PCR infection detection; and MDCK cells with infection detected using an immunofluorescence microscopic assay (IFA). Infections in CD-1 mice were detected by microscopy following staining with hematoxylin and eosin. Oocyst doses were enumerated by flow cytometry and delivered to mice and cell monolayers using standardized procedures. Each dose of oocysts was inoculated into up to nine replicates of 9 – 12 mice or 6 – 10 cell culture wells. Infectivity was expressed as the proportion of mice or cell monolayers that became infected at each oocyst challenge dose. Dose response curves were constructed by fitting a least squares regression to a plot of oocyst dose against a logistic transformation of the proportional infectivity and the 50% infectious dose (ID_{50}) values for each isolate in each infectivity model were calculated by solving the logistic regression for a logit value of zero. The majority of the comparisons used fresh oocysts that were less than 30 days old (post-shedding from propagation animal) but studies were also conducted with oocysts that had been exposed to environmental waters at ambient temperatures for >30 days.

The infectivity assays were also compared for oocysts exposed to disinfectants. Oocysts were exposed to ultraviolet light using a medium-pressure, collimated beam apparatus and inactivation was measured as the difference in ID_{50} of unexposed and UV-exposed oocysts. All UV exposures were performed at ambient temperature (typically 18 – 21°C). Oocysts were also exposed to ozone using batch, semi-batch, and single continuously stirred tank reactors. For ozone exposure, oocysts were suspended in phosphate buffer or untreated source waters at pH 7.4 – 8.3. Experiments were performed at 1, 5, and 15°C. Ozone inactivation was assessed by comparing ID_{50} values for exposed and non-exposed oocysts and by a logistic dose response.

CONCLUSIONS

The overall conclusion from this investigation is that in-vitro cell culture was equivalent to the "gold standard" mouse assay for measuring the infectivity of untreated *C. parvum* oocysts and should therefore be considered a practical alternative for assessing the potential of oocysts to cause infection. However, the high levels of variability displayed by mouse and cell culture methods indicated that infectivity and disinfection experiments should be limited to discerning relatively large differences. A cell culture-based assay was also demonstrated to be a practical method for evaluating the efficacy of some disinfectants such as UV light. However, further research is necessary to understand the response of oocysts to ozone when inactivation is assessed by cell culture methods. The specific conclusions of the project are:

1. There was good agreement between the CD-1 mouse assay and in-vitro cell culture for measuring the infectivity of fresh oocysts of five isolates of *C. parvum*. The general congruence was displayed in the shape of the dose response curves, the range of ID_{50} values, and the levels of variability.
2. Of the three cell culture assays used, infectivity results generated with the HCT-8/RT-PCR method were in closest agreement with results from the CD-1 mouse assay (Figure ES.1).
3. The ID_{50} values for the five *C. parvum* isolates in this study ranged from 13 to 629 for the various cell and mouse models. The mean and range of the ID_{50} values generated by the CD-1 mouse, HCT-8/RT-PCR cell culture, and Caco-2/RT-PCR cell culture assays, for all five isolates of *C. parvum* combined, were within the previously published 80% prediction limits of the Iowa isolate ID_{50} (Figure ES.2).
4. The wide variation in ID_{50} values for each isolate between the various assays makes it difficult to assign significance to the relative infectivity of different isolates of *C. parvum*. Nevertheless, the ID_{50} values obtained for the Iowa isolate were in general agreement with previous animal and human research (Table ES.1) and the TAMU and Moredun isolates appeared to be more infectious than other isolates.

5. The average standard deviation in proportional infectivity across all infectivity methods was 11.7%. Standard power analysis demonstrated that with three experimental replicates (10 mice or 10 wells of cell culture per replicate) the minimum difference in infectivity that could be detected was 23.7%. Reducing the detectable difference to 16% would require six replicates of 10 mice or 10 cell monolayers each (equivalent to a total of 60 mice or wells of cell culture) at each oocyst dose.
6. Infectivity differences between isolates may represent genuine differences in the levels of infectivity or they may simply be a reflection of combined sources of variability in the various assays.
7. Isolates of *C. hominis* were infectious in HCT-8 cell culture but did not infect mice.
8. Similar results were obtained with CD-1 mice and HCT-8 cell culture when measuring the infectivity of oocysts that had been exposed to an environmental water sample for 35 days.
9. There was very good agreement between the HCT-8 cell culture and CD-1 mouse assays for measuring UV inactivation of *C. parvum* (Figure ES.3). Combined data demonstrated that 2-$\log_{10}$ (99%) inactivation was achieved with a medium-pressure UV dosage of 5.6 mJ/cm^2.
10. The shapes of the ozone inactivation curves were generally the same for mouse and cell culture derived data although the CD-1 mouse assay typically generated 0.5 to 1-$\log_{10}$ higher levels of inactivation than the cell culture assay.
11. There was a pronounced stimulatory response in oocysts exposed to ozone below doses of 20 mg.min/L when assayed by HCT-8 cell culture. Sodium hypochlorite is a well known stimulant of oocysts and ozone appears to have similar properties when oocysts are inoculated into cell culture directly.
12. Up to 1.97-$\log_{10}$ inactivation was achieved in a single continuously stirred tank reactor with an ozone CT of 31.2 mg.min/L when infectivity was measured using CD-1 mice.
13. Using a batch reactor, at an ozone CT of 16 mg.min/L, 2 – 3 $\log_{10}$ inactivation was achieved based on data derived from the CD-1 mouse and HCT-8 cell culture assays.
14. In a semi-batch reactor, 3-$\log_{10}$ inactivation was achieved with ozone CTs of 14.9 and 42.3 mg.min/L at 15°C and 5°C, respectively, using CD-1 mice to measure infectivity.
15. Exposure to ozone at 1°C resulted in 3-$\log_{10}$ (99.9%) inactivation of the Iowa isolate of *C. parvum* with a CT of 66.2 mg.min/L as measured using the HCT-8 cell culture/RT-PCR assay.
16. The highest levels of ozone inactivation measured in this study were generally conservative relative to much of the published literature on ozone disinfection of *C. parvum* and comparison of all of the ozone disinfection data generated by this study indicated that overall the use of HCT-8 cell culture underestimated the level of inactivation relative to the CD-1 mouse assay (Figure ES.4).

Table ES.1
Comparison of ID_{50} values for the Iowa isolate of *Cryptosporidium parvum*

Infectivity model	ID50 (95% CI)	Reference
CD-1 mice	347 (128 – 942)	This study
HCT-8 cells	106 (52 – 154)	This study
CD-1 mice	60	Finch et al., 1993
CD-1 mice	87	Korich et al., 2000a
CD-1 mice	100 (84 – 119)	Korich et al., 2000b
Humans	132	Dupont et al., 1995
Humans	75 (34 – 114)	Okhuysen et al., 1999

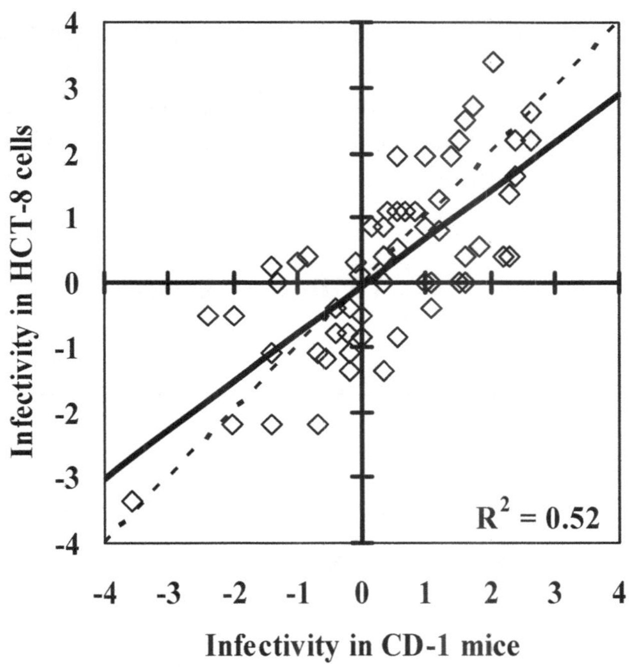

Figure ES.1 Comparison between HCT-8 cell culture and CD-1 mice for measuring infectivity of five *C. parvum* isolates (data for all isolates combined). The dashed line indicates the ideal correlation.

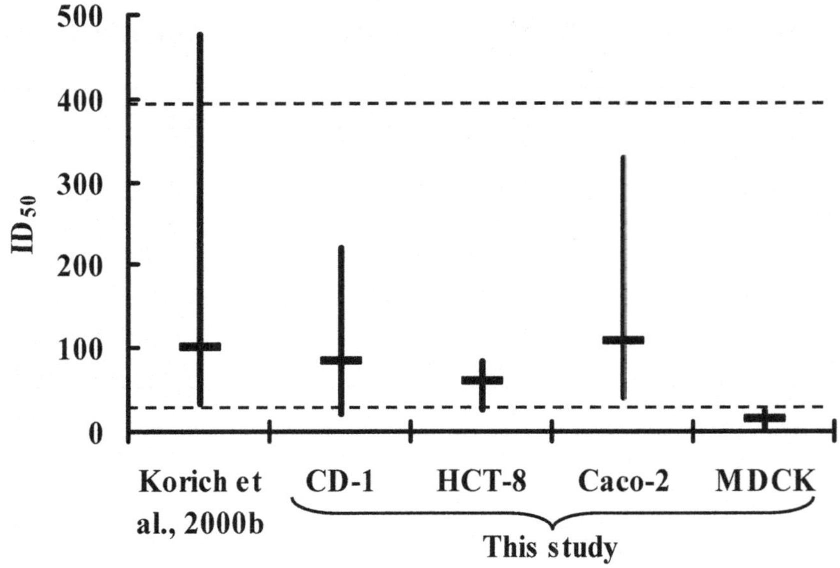

Figure ES.2 Average ID_{50} values (horizontal bars) and ranges (vertical bars) for the five isolates of *C. parvum* used in this study with each infectivity model compared to previously published data derived from 22 separate dose response determinations for just the Iowa isolate (Korich et al., 2000b). The horizontal dashed lines indicate the 80% prediction interval of the Korich data for the Iowa isolate.

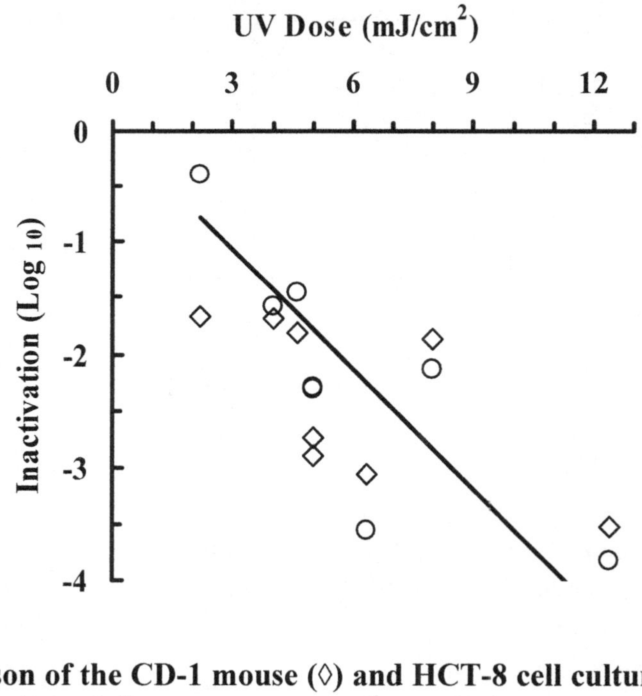

Figure ES.3 Comparison of the CD-1 mouse (◊) and HCT-8 cell culture/RT-PCR (○) assays for measuring inactivation of *C. parvum* oocysts by medium-pressure UV light. Regression lines forced through the origin were identical.

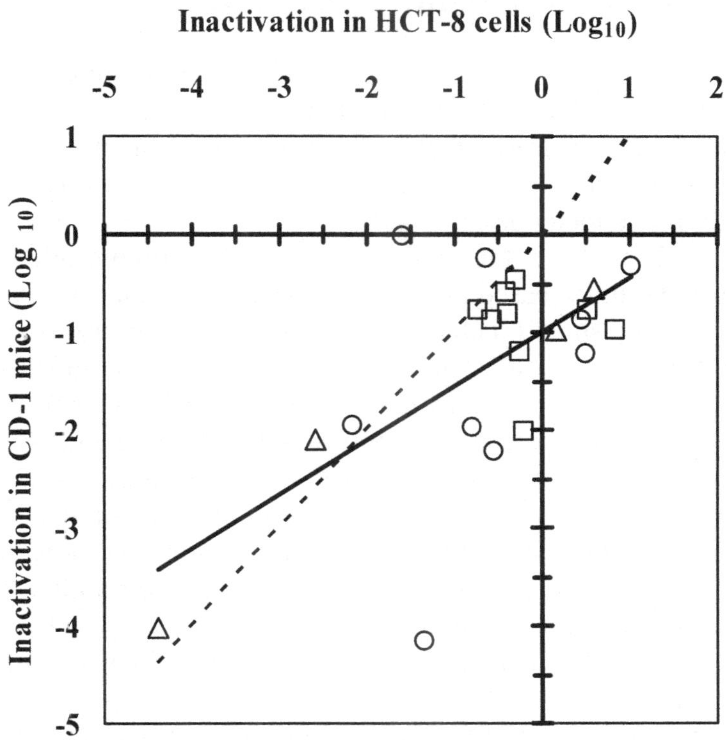

Figure ES.4 Comparison between HCT-8 cell culture/RT-PCR and CD-1 mouse assays for all ozone disinfection experiments that involved direct comparisons. Inactivation data were generated at 1°C, 5°C, and 15°C using batch (Δ), continuously stirred tank (□), and semi-batch (○) reactors. The dashed line represents the ideal correlation.

RECOMMENDATIONS AND FUTURE RESEARCH

1. The water industry should adopt in-vitro cell culture as a routine method for measuring the infectivity of waterborne *C. parvum* and *C. hominis* oocysts. This project has demonstrated that cell culture has equivalency with the standard CD-1 mouse assay and cell culture assays can be applied to oocysts recovered from water using approved methods.
2. There needs to be a thorough, robust, and well-controlled study to compare the various cell culture-based assays for measuring *C. parvum* and *C. hominis* infectivity. This evaluation should include inter-laboratory comparisons and round-robin testing.
3. Cell culture-based assays should be used to assess the infectivity of a range of fresh *C. hominis* isolates. These isolates should either be recently recovered from infected humans or propagated using the only suitable animal model (gnotobiotic piglet).
4. Cell culture-based assays should be used for assessing the efficacy of selected disinfectants, such as ultraviolet light and other irradiation methods.
5. Modifications to cell culture growth conditions and infection detection methods should be evaluated to determine whether assay variability can be reduced. Efforts should also be made to automate all or parts of cell culture assays so that a large number of replicate analyses can be performed, which will maximize the confidence in the resultant data.
6. Further research should be conducted on the response of oocysts to oxidants such as ozone and chlorine when infection is measured in cell culture. The kinetics and mechanisms involved in the stimulatory response observed at the lower ozone CTs in this project need to be more fully understood.
7. In-vitro cell culture methods need to be adapted so that they represent a better model of in-vivo systems. These adaptations should involve processing oocysts to simulate passage through the stomach prior to inoculation of cell monolayers. Incubation in bile salts (sodium taurocholate) has been demonstrated to be beneficial but additional treatments may be necessary.
8. The use of cell culture to measure inactivation by the application of sequential disinfectants should be assessed. It is likely that the stimulatory response will be eliminated by sequential disinfection.
9. The response of *C. hominis* oocysts to disinfection should be assessed in cell culture.
10. There should be more research aimed at developing a cell culture-based method for propagating oocysts of *C. parvum* and *C. hominis*. Developments in the methodology that may increase the likelihood of success include low-shear rotating wall culture vessels, scaled-up culture volumes, serum-free culture media, and medium supplements. A successful method would eliminate the reliance on calves for propagating the large numbers of *C. parvum* oocysts needed for experimental purposes.

CHAPTER 1
INTRODUCTION

Protozoan parasites of the *Cryptosporidium* genus are very common in many animal species including mammals, marsupials, reptiles, birds, and fish (Fayer et al., 1997; Fayer et al., 2000). Oocysts of various species are frequently detected in untreated surface waters (rivers and lakes) and have also been detected in groundwater and treated drinking water. For many years *Cryptosporidium parvum* was thought to be the only pathogen capable of causing infection in humans and so it has been the most extensively studied species. The organism is an obligate intracellular parasite that is excreted and exists in the environment as a dormant resistant oocyst. Oocysts are resistant to chlorine disinfection at the concentrations typically applied during drinking water treatment and there is no Food and Drug Administration approved treatment for cryptosporidiosis. It has been reported that, endemic levels of infection in developing regions of the world and the resulting persistent diarrhea in young children may lead to long-lasting impairment in growth and cognitive function (Guerrant et al., 1999). Although the disease is usually self-limiting in otherwise healthy patients, it can be fatal for those with weakened immune systems. Consequently, *C. parvum* presents a serious threat to public health and is a significant concern for the water industry.

BIOLOGY AND GENETICS OF *CRYPTOSPORIDIUM*

The genus *Cryptosporidium* contains at least eight recognized species that infect a variety of vertebrates. The organisms are coccidian parasites placed within the Phylum Apicomplexa. They are obligate, intracellular parasites that infect the epithelial cells lining the luminal surfaces of the digestive and respiratory tracts of a wide variety of animal hosts. Some species are restricted to a narrow range of host animals, e.g., *Cryptosporidium nasorum* in fish and *Cryptosporidium serpentis* in reptiles, while others such as *Cryptosporidium parvum* occur in a wide range of mammals. It is now recognized that *C. parvum* is actually a species complex; genotype 1 or anthroponotic oocysts have been isolated almost exclusively from humans while genotype 2 or zoonotic isolates have been detected in a wide range of mammals including humans. Based on genetic, physiological, and host animal differences, and the apparent lack of recombination between the two genotypes, they have recently been classified as separate species. The zoonotic genotype retained the name *C. parvum* while the anthroponotic genotype was named *C. hominis* (Morgan-Ryan et al., 2002). *C. parvum* and *C. hominis* oocysts (or organisms morphologically indistinguishable from *C. parvum* and *C. hominis*) have been reported in at least 152 species of mammals including humans (Fayer et al., 2000). A detailed review on the occurrence of *C. parvum* in animals, including farm, domestic, and wild animals, has been provided previously (Rochelle et al., 2001a). For many years it was thought that *C. parvum* (and by extension *C. hominis*) was the only species of *Cryptosporidium* that was capable of causing infection in humans. However, other species have also been detected in immune-compromised individuals: *C. canis*, *C. felis*, *C. meleagridis*, and *C. muris* (Fayer et al., 2001; Gatei et al., 2002; Morgan et al., 2000; Morgan-Ryan et al., 2002; Pedraza-Diaz et al., 2001; Pieniazek et al., 1999; Xiao et al., 2001). However, most cases of human cryptosporidiosis are attributed to *C. parvum* and *C. hominis*. Infections in humans may be asymptomatic but more frequently result in a variety of self-limiting acute enteric symptoms characterized by profuse diarrhea. There are no

Food and Drug Administration approved treatments for infected individuals and persistent infection in severely immune-compromised patients can contribute to mortality.

The life cycle of *C. parvum* features both sexual and asexual stages, thin-walled oocysts that reinfect the host leading to considerable biological amplification, and thick-walled oocysts that are excreted into the environment (Figure 1.1). Although dormant, the thick-walled oocyst stage of the parasite allows it to survive for prolonged periods and imparts a high level of resistance to most chemical disinfectants relative to many other pathogens. Unlike other apicomplexan parasites, the life cycle does not include a sporulation stage outside of the host; oocysts are infectious immediately upon shedding from an infected animal. Infection is acquired by ingestion of oocysts. Motile, short-lived sporozoites emerge from the oocyst and attach to intestinal epithelial cells inducing fusion of the microvilli; the parasite becomes engulfed by a double membrane comprised of host cellular material. A feeder organelle between the parasite and the host cell is the probable route of nutrients for the trophozoite. Trophozoites undergo asexual replication to produce 4 – 8 merozoites which are released into the intestinal lumen. The merozoites infect new cells and undergo additional rounds of merogony. Merozoites can also develop into macro- or microgametocytes. Microgametogenesis involves several rounds of replication followed by the release of numerous microgametes into the intestinal lumen. The microgametes fertilize macrogametes that are still attached to the intestinal epithelial cells. The resulting zygote undergoes sporogony and the resulting oocysts are either excreted in feces (thick-walled) or reinfect the host (thin-walled). A more detailed description of the life cycle of *Cryptosporidium* spp. has been provided previously (Fayer et al., 1997). A recent report described the observation of previously unrecognized extracellular stages in cell culture and animal infections with *C. andersonii* and *C. parvum* (Hijjawi et al., 2002). These stages (not represented in Figure 1.1) were present in large numbers and some of them were seen to undergo syzgy (state during sexual reproduction of some gregarines in which two or more life cycle stages connect end to end). The authors reported that the presence and characteristics of these extracellular developmental stages in the life cycle of *Cryptosporidium* provided support to the assertion that the genus is more closely related to the gregarine group of apicomplexans rather than the coccidia, a claim originally based on phylogenetic analysis of small subunit ribosomal RNA (SSU rRNA) sequences (Carreno et al., 1999).

The genetic differences between *C. parvum* (previously genotype 2 *C. parvum*) and *C. hominis* (previously genotype 1 *C. parvum*) have been demonstrated with many molecular markers including SSU rRNA, β-tubulin, and heat shock protein 70 (*hsp70*) genes (Figure 1.2; reviewed by Rochelle, 2001). The presence of two virus-like double-stranded RNA molecules has been confirmed in both *C. parvum* and *C. hominis* and polymorphisms in both of these molecules differentiated between them but the molecules have not been detected in other species (Khramtsov *et al.*, 2000). There have been only two instances of *C. hominis* recovered from non-human, naturally infected hosts. An isolate from a captive rhesus monkey was identified as *C. hominis* by multilocus genotyping with four polymorphic loci (Spano et al., 1998) and tissue sections from a sea cow were positive for *C. hominis* using primers targeting the 18S rRNA and acetyl CoA synthetase genes (Morgan et al., 2000).

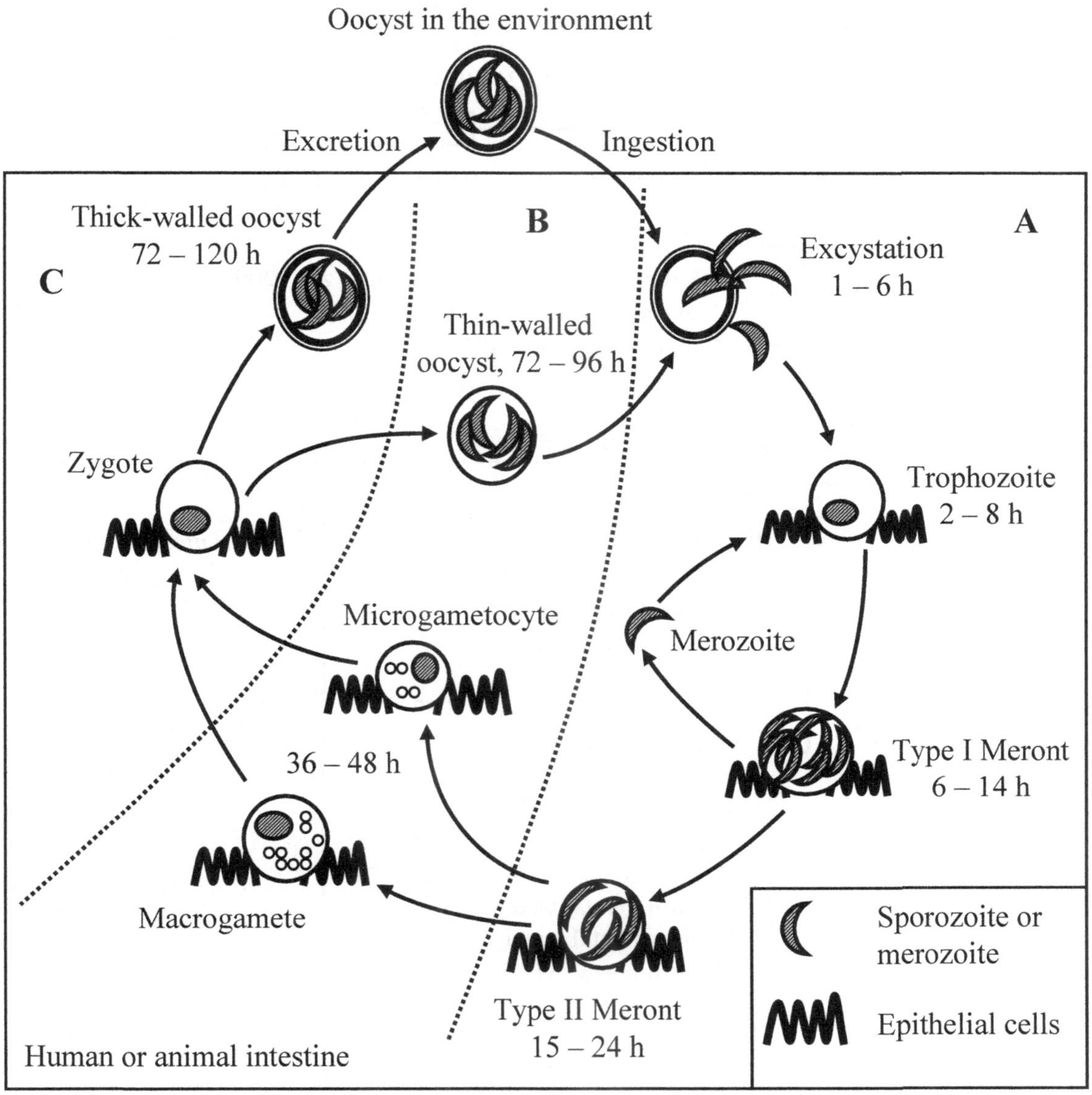

Figure 1.1 Life cycle of *Cryptosporidium parvum* indicating post-infection periods in hours at which different stages are detected, based on published reports (Buraud et al., 1991; Flanigan et al., 1991; Lawton et al., 1996; Rasmussen et al., 1993; Slifko et al., 1997). Life cycle diagram is adapted from Fayer et al. (1990). The life cycle stages in segment A (segments are delineated by dashed lines) have been well documented in animal and cell culture-based infections whereas those in segment B are not as strongly documented in cell cultures. Although the latter stages of the life cycle (segment C) have been reported in cell cultures, they only occur in high numbers (i.e., biological amplification) in human and animal infections.

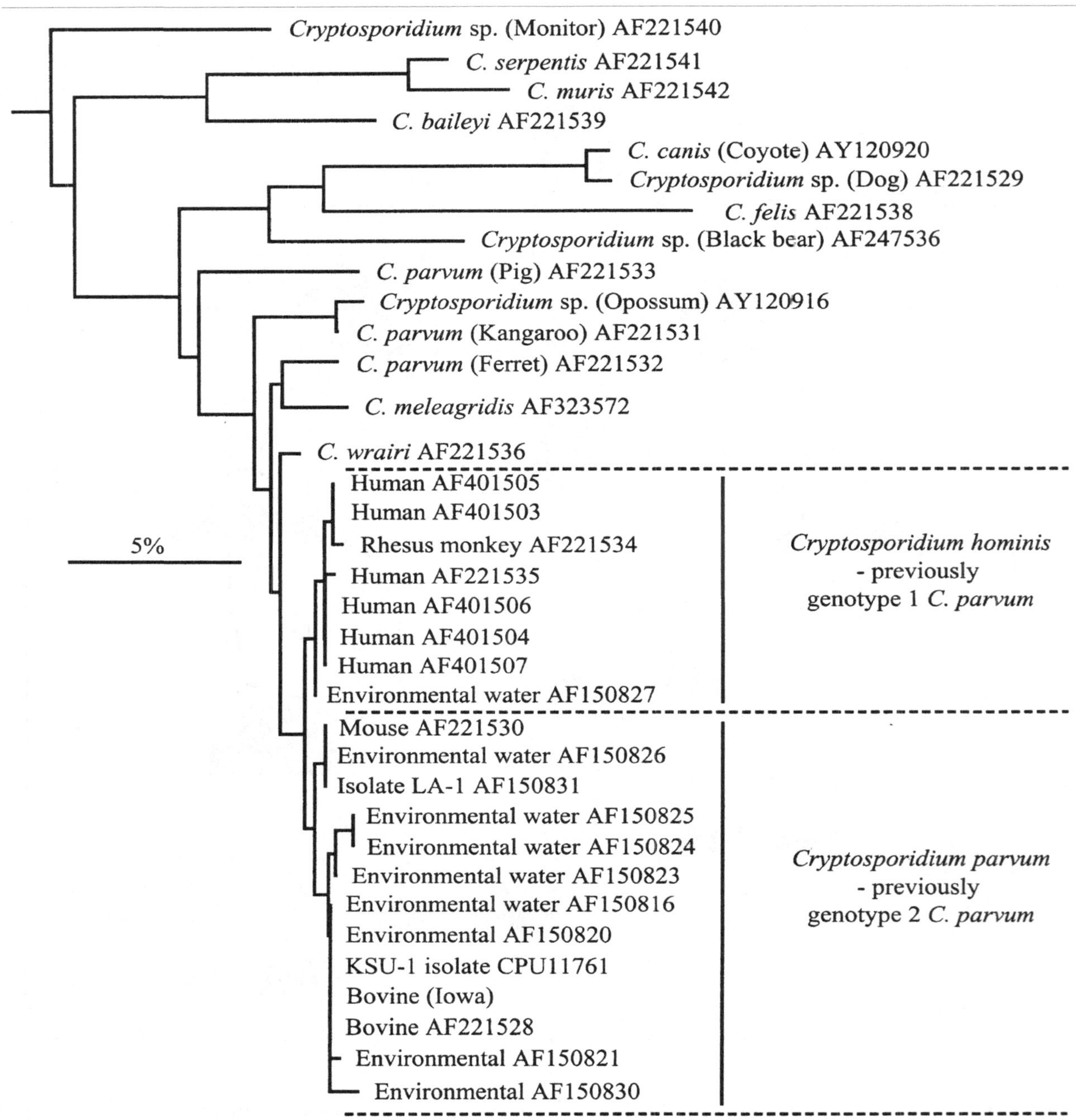

Figure 1.2 Heterogeneity within the 70 kD heat shock protein gene of the *Cryptosporidium* genus. The alignment covered the 361-bp fragment used for detection of infectivity in cell culture in this project (base positions 2423 to 2784, based on sequence CPU11761). Sequences were obtained from GenBank (accession numbers are displayed). The corresponding sequence of *Plasmodium falciparum* was used as the out-group. Distance calculations and tree construction were performed using the Kimura 2-parameter and neighbor joining methods, respectively. The scale bar indicates 5% sequence difference. (Source: Rochelle et al., 2004)

A compilation of genotyping data from 22 waterborne outbreaks of cryptosporidiosis demonstrated that 67% were caused by *C. hominis* and *C. parvum* was the causative agent in the remaining 33% (McLauchlin et al., 2000; Sulaiman et al., 1998). Genotyping techniques have also been applied to endemic infections. In one study using restriction fragment length polymorphism (RFLP) analysis of specific amplicons, 33% of isolates from human samples were characterized as *C. hominis* and 67% were *C. parvum* (Rochelle et al., 2001a). McLauchlin et al. (2000) reported similar results with 38% of human samples characterized as *C. hominis* and 62% as *C. parvum* (N = 1,705). Both species were identified in 0.4% of samples and *C. meleagridis* was also detected in 0.3% of these human samples. These authors also reported a marked geographic distribution of genotypes amongst sporadic cases of cryptosporidiosis and *C. hominis* was more prevalent in patients with a recent history of foreign travel. However, Sulaiman et al. (1998) reported a different trend. Out of 29 human samples, 72% and 28% were identified as *C. hominis* and *C. parvum*, respectively. Of 17 samples from HIV-infected patients, 88% were *C. hominis*. Apart from the inter-species genetic variation there is also considerable strain level variation in the genomic sequences within each species (Figure 1.2). For example, PCR-sequencing of 94 isolates targeting a microsatellite marker demonstrated that *C. hominis* can be divided into two sub-genotypes and *C. parvum* can be further differentiated into four sub-types (Caccio et al., 2000). Some of the sub-genotypes were restricted to specific geographic locations while others displayed a broad geographic distribution. In addition, Xiao et al. (1999) reported that *C. hominis* could be further divided into three, four, or five sub-genotypes based on the sequences of polythreonine region, 18S rRNA, or hsp70 genes, respectively.

CRYPTOSPORIDIUM IN DRINKING WATER

Since 1983 more than 55 documented outbreaks of cryptosporidiosis have been associated with either drinking water or recreational use of water worldwide (reviewed by Rochelle et al., 2001a and Fayer et al., 2000). The largest waterborne outbreak to date occurred in 1993 in Milwaukee with estimates of the affected population ranging from 15,000 to 400,000 individuals (Hunter and Syed, 2001; MacKenzie et al., 1994) and up to 100 deaths attributed to the outbreak. Over the last 10 years, the continued detection of *Cryptosporidium* oocysts in treated drinking water has led to boil water advisories affecting millions of people worldwide. A review of eight studies assessing the occurrence of *Cryptosporidium* spp. in untreated source waters in the United States demonstrated that the average proportion of river, lake, and well water samples that were contaminated with oocysts ranged from 9 to 100% (Rose et al., 1997). Water samples that were impacted by domestic and agricultural waste had oocyst concentrations as high as 5,800/L (Madore et al., 1987). A large survey of North America spanning 1988 – 1993 reported that 60.2% of samples (N = 347) were positive for *Cryptosporidium* oocysts (LeChevallier and Norton, 1995). A similar study in Canada demonstrated lower levels of contamination with oocysts detected in 6.1%, 4.5%, and 3.5% of raw sewage, raw water, and treated drinking water, respectively (Wallis et al., 1996). Additional studies have reported the occurrence of oocysts in 6% of stream samples in Wisconsin (Archer et al., 1995), 63% of river samples in Pennsylvania (States et al., 1997), and 13% of surface waters in New Zealand (Ionas et al., 1998). A large watershed survey detected oocysts in 11% of samples (N = 189) and 24% of first flush samples (N = 34) with extrapolated oocyst concentrations up to 417/L following storm events (Ferguson et al., 1998). In a relatively large outbreak of waterborne cryptosporidiosis in the U.K., a total of 360 cases of infection were confirmed and oocysts were detected at a concentration of 340 oocysts per 100 L in the reservoir serving the community. In addition, *Cryptosporidium* oocysts were detected in 37.5% (N = 32) of fecal samples collected from sheep grazing around the reservoir (CDR, 1999). A survey of 5,838 untreated source waters throughout the U.S. reported an average occurrence of 6.8% with a mean concentration of 0.067 oocysts/L, although the detection method had considerable variability (Messner and Wolpert, 2003). Correctly operating water treatment plants that utilize filtration usually remove oocysts from source water with high efficiency. However, oocysts have been detected in 3.8 – 40% of treated drinking water samples at concentrations up to 48 oocysts/100 L (Rose et al., 1997). A survey of treatment plants in Wisconsin detected oocysts in 4.2% (N = 72) of finished water samples (Archer et al., 1995).

Recent advances in molecular characterization of *Cryptosporidium* spp. have also been applied to the detection of oocysts in water. Out of a total of 29 storm water samples analyzed by a PCR-RFLP analysis targeting the SSU rRNA gene, 27 (93.1%) were positive for *Cryptosporidium* with 12 different genotypes represented (Xiao et al., 2000). None of the detected genotypes matched those typically found in human, farm animal, or domestic animal samples. However, four were identical or closely related to *C. baileyi*, and *Cryptosporidium* genotypes from opossums and snakes indicating that wildlife was the primary source of oocyst contamination of surface water during storms. The same method was also used to analyze untreated surface water and wastewater samples throughout the U.S. *Cryptosporidium* was detected in 45.5% of surface water samples (N = 55) and 24.5% of raw wastewater samples (N = 49; Xiao et al., 2001). The predominant genotypes in surface water matched the profiles of *C. parvum* and *C. hominis* while *C. andersoni* was most commonly detected in wastewater.

MEASURING INFECTIVITY OF *CRYPTOSPORIDIUM*

To fully assess the public health significance of waterborne *C. parvum* and *C. hominis*, it is necessary to evaluate the viability and/or infectivity of oocysts in environmental waters. Also, due to the resistance of oocysts to standard chlorine-based disinfectants and most chemotherapeutic agents, the use of such assays is becomingly increasingly important to evaluate the efficacy of alternative drugs and disinfectants.

There are a variety of in-vitro methods for determining the viability of oocysts such as in-vitro excystation, staining with vital dyes, and detection of mRNA by RT-PCR. A dye permeability assay combining 4',6'-diamidino-2-phenylindole/propidium iodide (DAPI/PI) was reported to be useful as an indicator of the potential viability and infectivity of oocysts (Jenkins et al., 1997) and viability indicated by the nucleic acid stains SYTO-9 and SYTO-59 correlated with mouse infectivity results (Belosevic et al., 1997). In addition, staining oocysts with SYTO-9 followed by flow cytometric detection was found to provide a good estimate of oocyst inactivation by ozone (Gyurek et al., 2001). However, it has been demonstrated that these staining methods do not always correlate with infectivity assays. A comparison of a mouse infectivity assay with four in-vitro viability assays (excystation and staining with DAPI/PI, SYTO-9 and SYTO-59) demonstrated that the in-vitro surrogate methods consistently overestimated oocyst viability and could not be used to accurately assess ozone disinfection of *C. parvum* (Bukhari et al., 2000). The authors concluded that neonatal mouse infectivity should be considered as the "gold standard" for evaluating oocyst disinfection. Earlier studies of UV exposure using surrogate measures of viability such as in-vitro excystation and uptake of vital dyes indicated that very high dosages of UV (8,700 mJ/cm^2) were necessary to achieve 2-log_{10} (99%) inactivation of *C. parvum* (Campbell et al., 1995). In contrast, infectivity assays have demonstrated that the parasites ability to infect is eliminated at relatively low doses of UV (2-log_{10} inactivation at 6 mJ/cm^2; Mofidi et al., 2001). Thus, UV-exposed oocysts appear able to excyst (a process that does not require DNA replication) but not to initiate an infection, a complex process requiring DNA replication and cellular differentiation. RT-PCR amplification of RNA targets has also been used to assess oocyst viability but the choice of RNA target strongly influences the results (Stinear et al., 1996; Widmer et al., 1999). Also, there has been no assessment of the utility of the method for measuring disinfection and only limited comparisons of the RT-PCR viability approach with animal infectivity have been performed. Consequently, assays that can determine the ability of oocysts to cause infection, rather than simply detecting some aspect of metabolism or presence of particular molecules, are essential. The three methods used to assess *C. parvum* infectivity are human volunteer studies, animal assays, and in-vitro cell culture.

Human Volunteer Studies

From the perspective of data accuracy and extrapolation, the ideal model for assessing the infectivity of a human pathogen would be a representative selection of *Homo sapiens*. However, human infectivity assays are not practical for use on a routine basis due to the difficulty in obtaining a sufficiently large cohort size, ethical concerns surrounding human testing, and the potentially serious and long-term health effects for study participants. A typical *C. parvum* disinfection experiment may utilize up to 240 animals or over 400 wells of cell culture. Obviously, conducting experimental infections on such a large group of humans is not feasible. The difficulties in performing human feeding trials are exemplified by the fact that all of the studies to date have been conducted by a single group of researchers over a 10 year period. Nevertheless, a great deal of useful information has been gleaned from the limited number of carefully controlled human feeding trials that have been performed. In one study, the 50% infectious dose (ID_{50}) for the Iowa isolate (bovine origin) of *C. parvum* in healthy human volunteers was 132 oocysts (N = 29) with some individuals becoming infected with 34 oocysts, the lowest challenge doses administered (DuPont et al., 1995). A follow-up investigation using the same volunteers 1 year later demonstrated that the initial exposure to oocysts did not protect against infection when challenged a second time with 500 oocysts (the 86% infectious dose; Okhuysen et al., 1998a). In a further study, the 100% infectious dose for the UCP isolate (also bovine origin) was between 5,000 and 10,000 oocysts (Okhuysen et al., 1998b). This study also demonstrated that administration of bovine anti-*Cryptosporidium* immunoglobulin to study subjects provided no protection against infection. A second study with the UCP isolate also generated a higher ID_{50} (Okhuysen et al., 1999). When using oocyst excretion as the measure of infection, a comparison of three isolates demonstrated an ID_{50} of 75 for the Iowa isolate (N = 29; 95% confidence interval [CI], 34 – 114), 125 for the TAMU isolate originally of equine origin (N = 14; 95% CI, 77 – 179), and 2,788 for the UCP isolate (N = 17; 95% CI, 0 – 11,214). Finally, the ID_{50} for the Moredun isolate (cervine origin) was approximately 300 – 375 oocysts (N = 16; Okhuysen et al., 2002). Tissue biopsies from subjects in these various experiments aimed at assessing ID_{50} values have also been used to elucidate the role of cytokines in controlling cryptosporidiosis. No intentional human feeding trials have been conducted with *C. hominis* although accidental infection was reported in three laboratory personnel who were caring for artificially infected calves (Akiyoshi et al., 2002).

Animal Models

Artificial infection of calves is the most common method for propagating oocysts of *C. parvum* since the parasite undergoes considerable amplification in newborn cattle; an amplification of 50 to 150-fold is typical in artificially infected calves with a single animal shedding up to 2×10^{10} oocysts (M. Marshall, unpublished observations). A wide variety of animals, including hamsters, rats, macaques, pigs, lambs, and opossums have been used for *C. parvum* infectivity assays, but the most commonly used animal models are various strains of adult and suckling mice. The ID_{50} of the Iowa isolate in CD-1 mice has been reported as 60 – 87 oocysts (Finch et al., 1993; Korich et al., 2000a). However, natural lot-to-lot variability of the Iowa isolate was demonstrated by a review of 22 dose-response studies in CD-1 mice over a three year period with oocysts obtained from 10 calves and assayed in a total of 1,700 mice. The mean ID_{50} was 100 oocysts (95% CI, 84 – 119) with an ID_{50} range of 33 – 476 oocysts (Korich et al., 2000b). Various other mouse models have been used to measure the infectivity of *C. parvum* oocysts, with ID_{50} values ranging from 60 to 1,000 oocysts (Lindsay, 1997). Gamma interferon knockout (GKO) mice can be infected with as few as 10 oocysts of *C. parvum* (Griffiths 1998) and the Moredun isolate was shown to be highly infectious in GKO mice with infections resulting from inoculation with a single oocyst, based on serial dilution of a hemacytometer enumerated stock preparation (Okhuysen et al., 2002). It was also reported that the ID_{50} of *C. parvum* in NMRI mice was less than 10 oocysts, based on inoculations with serial dilutions of an enumerated stock and flow cytometry to detect shed oocysts (Delaunay et al., 2000). An isolate of *C. meleagridis*, originally isolated from an infected child, was successfully propagated in gnotobiotic piglets and passaged in mice, chickens, turkey poults, and calves indicating that *C. meleagridis* was readily transmissible between and among mammalian and avian animal species (Akiyoshi et al., 2003).

Although infectivity in mice is generally considered to be the gold standard for measuring *C. parvum* infectivity, there are drawbacks to the approach. The use of animals in scientific research raises ethical concerns, and animal-based assays are expensive and time-consuming and have significant hidden costs, such as the maintenance of accredited facilities and license fees. A further disadvantage has been the inability of mouse infectivity models, including GKO mice, to support infection of *C. hominis* (Peng et al., 1997; Widmer et al., 2000a). *C. hominis* has been propagated to a limited extent in gnotobiotic piglets and has demonstrated infection in calves (Akiyoshi et al., 2002). However, no routine infectivity assay is yet available for testing the infectivity or response to disinfectants of *C. hominis*.

In-Vitro Cell Culture

A variety of mammalian cell lines have been reported to support infection by *C. parvum* and *C. hominis* (Table 1.1). Complete development is defined as the de-novo production of oocysts as a result of the parasite undergoing its full life cycle in cell culture and was first reported by Current and Haynes (1984). Sporozoites were inoculated onto human fetal lung cells and infection was detected by Nomarski interference contrast microscopy. Macrogametes and microgametocytes (sexual stages) were present within 48 hours of inoculation and new oocysts were detected 72 hours after inoculation. Complete development was also reported in Caco-2, RL95-2, BFTE, THP-1 and MDBK cells (Buraud et al., 1991; Rasmussen et al., 1993; Yang et al., 1996; Lawton et al., 1996). Oocysts were also observed in monolayers of HCT-8 and HT-29 cells that were inoculated with excsyted sporozoites (Maillot et al., 1997). However, routine procedures for oocyst propagation in cell culture have not been developed; the number of oocysts produced is low and always less than the number of oocysts in the original inoculum. The failure of *C. parvum* to propagate in cell culture is not yet understood but may be related to deficiencies in the culture medium or problems with the host cells resulting in the inability of the parasite to efficiently progress beyond the asexual stage of the life cycle. Invasion and infection may result in apoptosis (programmed cell death) in the host monolayer (Widmer et al., 2000b) limiting the prospects for long term parasite development. One group of researchers reported development of the complete life cycle of *C. parvum* and *C. hominis* in HCT-8 cells and maintenance of the infection for 25 days by repeated subculturing (Hijjawi et al., 2001). This group also described the complete development of *C. andersoni* in HCT-8 cells and the presence of previously undescribed extracellular life cycle stages (Hijjawi et al., 2002).

In a comparison of a variety of cell lines for supporting in-vitro growth of *C. parvum*, infection in HCT-8 cells generated approximately twice as many intracellular life cycle stages compared to MDBK, MDCK, or Caco-2 cells (Upton et al., 1994). However, other investigators reported that there was no difference in the level of infection supported by Caco-2, HCT-8 or HT29 cell lines (Maillot et al., 1997) or when comparing HCT-8 and MDCK cells (You et al., 1996). Although the great majority of in-vitro infections have been with *C. parvum* or *C. hominis*, other species have also been studied. For example, MDBK cells supported infection with *C. meleagridis* (Akiyoshi et al., 2003).

Different cell lines require particular types of media and additives for optimum growth, although optimum growth conditions for the host cells may not be conducive to maximum parasite development. Standard cell culture media contain all of the essential nutrients for cell growth and are typically supplemented with fetal bovine serum (FBS) at concentrations ranging from 5% to 20%. A variety of media additives may also be added depending on the application and the preference or experience of the researchers. These include: antibiotics to suppress growth of bacterial and fungal contaminants (typically penicillin, streptomycin, kanamycin, amphotericin), HEPES buffer, glutamine, glucose, vitamins, and insulin. Detailed reviews of in-vitro cell culture approaches and media formulations have been published previously (Arrowood, 2002; Rochelle and De Leon, 2001; Upton, 1997).

A variety of infectivity assays have been developed for *C. parvum* using various cell lines, assay formats, and detection methods (Table 1.2). Any of these assays could be adapted to a variety of applications. The ELISA and chemiluminescence immunoassays in 96-well formats have been particularly useful for large scale screening of potential anticryptosporidial agents (Woods et al., 1995; You et al., 1996). The microscopic detection methods provide readily enumerated infectivity results because the actual life stages are visualized. However, such procedures can be time consuming. Molecular-based infection detection methods utilizing PCR to amplify either DNA or mRNA have also been developed (Di Giovanni et al., 1999; Rochelle et al., 1997). These techniques are highly specific and sensitive and, as with the enzyme-linked methods described above, can be used to screen a large number of samples simultaneously. An assay using RT-PCR to amplify *C. parvum*-specific mRNA from a region of the heat shock protein gene (hsp70) was used to detect infection in Caco-2 and HCT-8 cells with as few as 10 oocysts (Rochelle et al., 1997). Although PCR and RT-PCR methods can be used for relative quantitation, which is useful for inactivation studies, it is difficult to generate the absolute quantitative data provided by microscopic methods. Consequently, a colorimetric in-situ hybridization assay (CISH) was developed to provide the specificity of molecular-based methods with the easy enumeration of microscopic methods (Rochelle et al., 2001b). RT-PCR detection of infection on HCT-8 cells was used to demonstrate that oocysts recovered from environmental water samples by immunomagnetic separation and by USEPA Method 1622 retained their infectivity (Rochelle et al., 1999a). Also, using HCT-8 cells and PCR detection, 4.9% of raw water samples and 7.4% of filter backwash samples were found to contain infectious *C. parvum* (Di Giovanni et al., 1999). The sensitivity of the latter assay was reported to be less than five infectious oocysts. A sensitivity of a single infectious oocyst was reported for an assay using IFA to detect infection in HCT-8 cells (Slifko et al., 1997). However, standardization is necessary to ensure these reported sensitivities are obtained on a routine basis. *C. parvum* oocysts demonstrate considerable variation in infectivity, even with a single isolate prepared at different times. It has been established by many researchers that oocysts lose infectivity as they age and greater than 10-fold differences in cell culture infectivity have been reported for different lots of the Iowa isolate of the same age (Slifko et al., 1999).

Table 1.1
Cell lines that support the in-vitro growth of *C. parvum*

Cell line	Origin of cells	Reference
-	Mouse peritoneal macrophages	Martinez et al., 1992
A-549	Human lung carcinoma	Giacometti et al., 2001
BALB/3T3	BALB/c mouse embryo	Upton et al., 1994
BFTE	Bovine fallopian tube epithelium	Forney et al., 1996
BHK	Baby hamster kidney	Naciri et al., 1986
BS-C-1	African green monkey kidney	Deng and Cliver, 1998
BT-549	Human breast infiltrating duct carcinoma	Upton et al., 1994
Caco-2	Human colonic adenocarcinoma	Buraud et al., 1991
H69	Human bile duct epithelium	Chen et al., 1998
HCT-8	Human ileocecal adenocarcinoma	Upton et al., 1994
HFL	Human fetal lung	Current and Haynes, 1984
Hs-700T	Human pelvic adenocarcinmoa	Upton et al., 1994
HT-1080	Human fibrosarcoma	Upton et al., 1994
HT-29	Human colonic adenocarcinoma	Flanigan et al., 1991
LGA	Rat small intestine carcinoma	Bonnin et al., 1990
LS-174T	Human colonic adenocarcinmoa	Upton et al., 1994
MDBK	Bovine kidney	Upton et al., 1994
MDCK	Canine kidney	Gut et al., 1991
RL95-2	Human endometrial carcinoma	Rasmussen et al., 1993
T84	Human colonic carcinoma	Adams et al., 1994
THP-1	Non-adherent human monocytes	Lawton et al., 1996

Table 1.2
Cell culture-based infectivity assays for *C. parvum*

Cell line	Assay format	Detection method	Reference
RL95-2	Thermanox coverslips	Giemsa staining	Rasmussen et al., 1993
MDBK	Glass coverslips	Interference microscopy	Upton et al., 1994
MDCK	96-well plates	Chemiluminescence immunoassay	You et al., 1996
MDCK	Chamber slides	Immunofluorescence	Arrowood et al., 1994
BFTE	Glass coverslips	Microscopy	Forney et al., 1996
Caco-2	Permeable membranes	Transmonolayer resistance	Griffiths et al., 1994
Caco-2	Chamber slides	RT-PCR on messenger RNA	Rochelle et al., 1997
HCT-8	96-well plates	ELISA	Woods et al., 1995
HCT-8	24-well plates	RT-PCR on messenger RNA	Rochelle et al., 1999a
HCT-8	Chamber slides	Immunofluorescence	Slifko et al., 1997
HCT-8	Chamber slides	In-situ hybridization	Rochelle et al., 2001b
HCT-8	96-well plates	PCR on DNA	Di Giovanni et al., 1999
BS-C-1	24-well plates	PCR on DNA	Deng and Cliver, 1998

DISINFECTION OF *CRYPTOSPORIDIUM PARVUM*

The primary disinfectants used throughout the water industry in the United States for inactivation of microorganisms in drinking water are chlorine and chloramines. However, standard chlorine-based disinfectants have little effect on *C. parvum* oocysts at concentrations typically applied in drinking water treatment plants. In fact, incubation in 10% bleach (0.5% sodium hypochlorite) is a routine pretreatment used to enhance excystation for cell culture (Upton et al., 1994). Some estimates of the chlorine CT (product of disinfectant concentration multiplied by contact time) for *C. parvum* oocysts are as high as 7,200 mg.min/L for >99% inactivation (Korich et al., 1990) compared to a typical value of 0.04 mg.min/L for *Escherichia coli*. Although chlorine dioxide is a more effective disinfectant than free chlorine, CT values of 75 – 1,000 mg.min/L were necessary to achieve 2-$\log_{10}$ (99%) inactivation (Chauret et al., 2001), values that are much higher than concentrations typically used in drinking water treatment. A compilation of selected published data, without considering experimental conditions, indicates that chlorine dioxide CT values of 143.8 and 633 mg.min/L are required to achieve 2-$\log_{10}$ and 3-$\log_{10}$ inactivation of *C. parvum*, respectively (Figure 1.3). Therefore, alternative disinfectants such as ozone, ultraviolet (UV) light, pulsed white light, ionizing radiation, and bromine have been investigated. Possibly the most promising of these alternatives is UV irradiation. UV has been used for many years to disinfect water intended for pharmaceutical, medical, and food preparation purposes and it is used routinely in some parts of Europe for disinfection of drinking water. The availability of alternative disinfectants that need to be evaluated emphasizes the requirement for routine, reliable, practical, and accurate infectivity assays.

Recent studies have demonstrated that UV light appears to be effective for inactivation of *C. parvum* oocysts. Studies using mouse assays to measure oocyst infectivity and inactivation reported 3.9-$\log_{10}$ (99.99%) inactivation at a medium-pressure UV dosage of 19 mJ/cm^2

(Bukhari et al., 1999) and 2- to 3-$\log_{10}$ (99% – 99.9%) inactivation at 10 to 25 mJ/cm^2 of medium-pressure UV, respectively (Craik et al., 2001). An in-vitro cell culture-based method combined with a fluorescent antibody was used to demonstrate 3-$\log_{10}$ (99.9%) inactivation at a dosage of 3 mJ/cm^2 of low pressure UV (Shin et al., 2001) while a cell culture procedure combined with a reverse transcriptase-polymerase chain reaction (RT-PCR) assay demonstrated that polychromatic UV doses of 3 mJ/cm^2 and 6 mJ/cm^2 provided an average of 1-$\log_{10}$ (90%) and 2-$\log_{10}$ (99%) inactivation of the Iowa isolate of *C. parvum*, respectively (Mofidi et al., 2001). This latter study also demonstrated that there was no significant difference between medium-pressure and pulsed-UV lamps for inactivation of *C. parvum*.

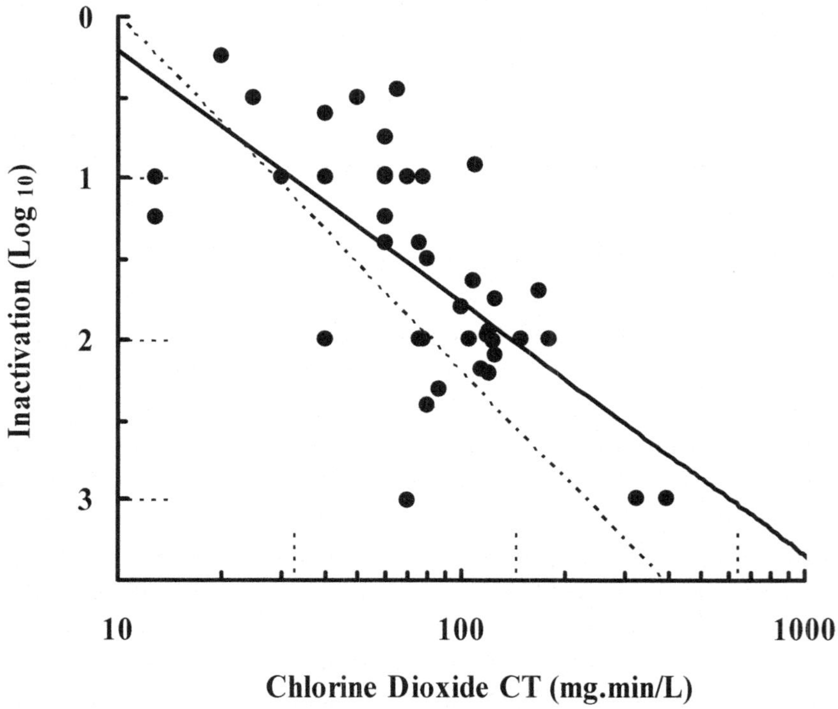

Figure 1.3 Chlorine dioxide inactivation of *Cryptosporidium parvum* oocysts. Data was extracted from the following references without regard to the experimental design or conditions: Chauret et al., 2001; Di Giovanni et al., 1999; Finch et al., 1995; Owens et al., 1999; Ruffell et al., 2000. The solid line is the regression through the plotted data. The dashed line is the average regression through cell culture-derived data for three lots of oocysts (Owens et al., 1999).

Ozone Inactivation Studies

Ozone is highly reactive, with a comparative oxidizing strength of 2.07 volts relative to hydrogen, compared to 1.36 and 1.28 volts for chlorine and chlorine dioxide, respectively. This high reactivity results in rapid oxidation of organic material and thus leads to effective inactivation of microorganisms through a process that involves disruption of cell surfaces and walls. Ozone has been used to disinfect water for over 100 years but was supplanted by chlorine before it gained widespread and routine use. A detailed history of the application of ozone for the treatment of drinking water around the world was provided by Brink et al. (1991) and White (1999). Over the last 20 years, the emergence of *Cryptosporidium* as a waterborne pathogen of significant concern and the realization that oocysts are relatively resistant to standard chlorine disinfection practices has been one of the driving forces behind the increased interest in ozone as a disinfectant for water.

There have been over 40 reports describing the efficacy of ozone for inactivation of *Cryptosporidium* oocysts. The experimental design of many of these studies and their respective results have been reviewed in varying levels of detail by previously published reports (Clark et al., 2002; Haas et al., 2001; Oppenheimer et al., 2000). Select examples of inactivation data from two of these studies are displayed in Figures 1.4 and 1.5. These figures are only intended to be illustrative of the type of data available, the variability in the data, and the affect of temperature on ozone inactivation of *C. parvum*. Consequently, simple linear regressions lines were plotted through the data with no attempt to model the most appropriate inactivation curves. Based on these linear regressions, the Oppenheimer data obtained with a single continuously stirred tank reactor (CSTR) demonstrates that CT values of 2, 5.4, and 38.4 mg.min/L are required to achieve 2-$\log_{10}$ inactivation at 23°C, 14°C, and 3°C, respectively (Figure 1.4). A more thorough analysis of CT requirements extracted from the authors' report (Oppenheimer et al., 2000) is presented in Table 1.3. The major findings of this report were summarized as follows:

1. Ozone was effective for inactivating *C. parvum* but the efficiency of inactivation may be affected by the pH (increased inactivation at pH $\leq$ 6.5) and turbidity of the water.
2. The CT product increased by a factor of 4.5 for each 10°C decrease in water temperature (compared to a factor of 2 for *Giardia duodenalis*).
3. Ozone inactivation data was adequately described by the linear Chick-Watson model.
4. The application of chlorine or chloramines following ozone may result in synergistic disinfection.
5. Ozone disinfection of *C. parvum* may require elevated CT values that result in the formation of unacceptably high levels of bromate (the most significant ozone disinfection byproduct).

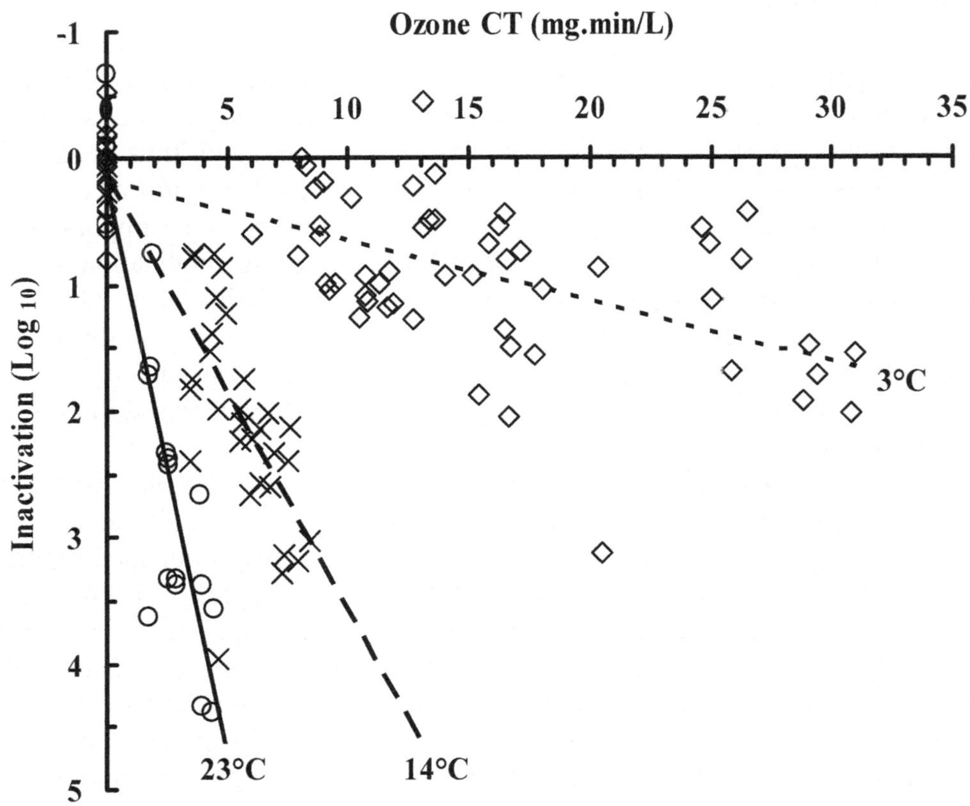

Figure 1.4 Inactivation of *Cryptosporidium parvum* by ozone in a single continuously stirred tank reactor. The figure displays a selected subset of data extracted from Oppenheimer et al. (2000). Oocysts were suspended in environmental source waters at 23°C (○), 14°C (×), and 3°C (◊).

Table 1.3
Summary of ozone CT requirements for inactivation of *C. parvum*
(from Oppenheimer et al., 2000)

		CT required for indicated inactivation:			
Temperature	Percentile	0.5-log_{10}	1-log_{10}	2-log_{10}	3-log_{10}
3°C	10	5.5	11	22	33
3°C	50	7.8	16	31	47
3°C	90	13	26	52	79
25°C	10	0.2	0.4	0.8	1.2
25°C	50	0.3	0.5	1.1	1.6
25°C	90	0.5	0.9	1.8	2.8

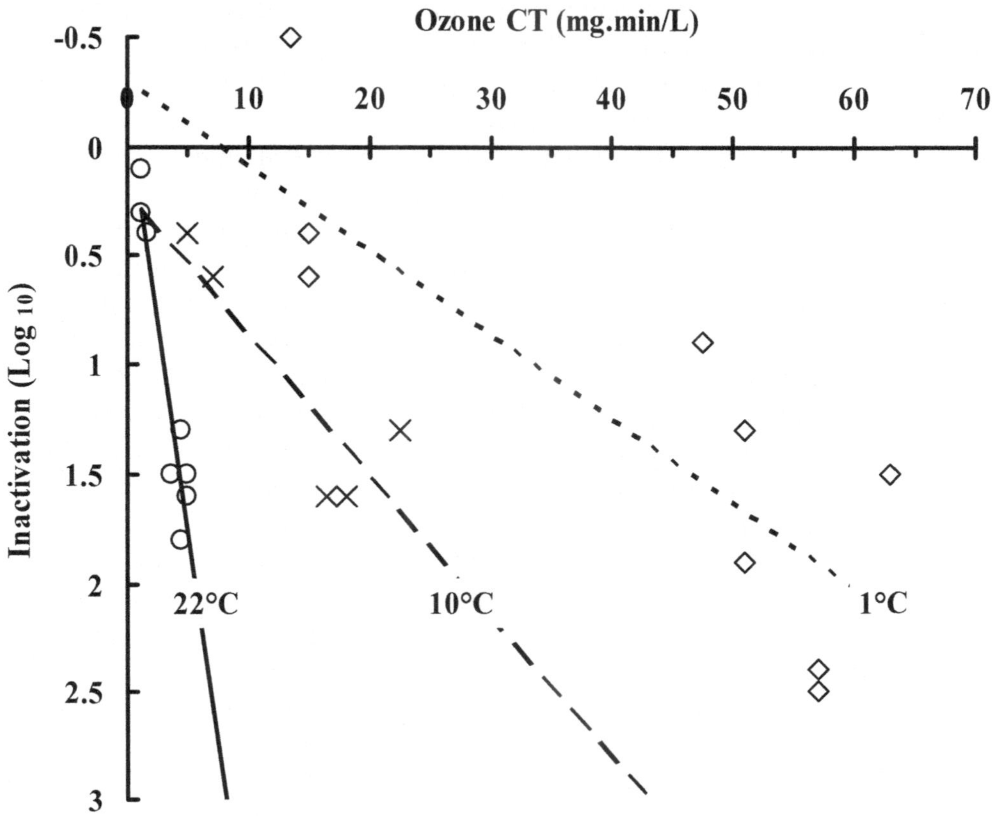

Figure 1.5 Inactivation of *Cryptosporidium parvum* in a batch reactor. Data was extracted from Li et al. (2001) for oocysts suspended in 0.05 M phosphate buffer at 22°C (○), 10°C (×), and 1°C (◊).

In the second example, Li et al. (2001) studied inactivation in a batch reactor with oocysts (Iowa isolate) suspended in 0.05 M phosphate buffer. Linear regressions of the data indicated that CT values of 5.62, 27.6, and 59.4 mg.min/L were required to achieve 2-$\log_{10}$ inactivation at 22°C, 10°C, and 1°C, respectively (Figure 1.5).

There is considerable variation in the published ozone inactivation data due in part to the different experimental methods used in the various studies (differences in reactor types, test matrices, methods used to assess inactivation, mathematical and statistical treatment of data) but also due to the variability inherent in biological assays, particularly those involving an obligate intracellular parasite such as *Cryptosporidium*.

A recent critical review of the physical and chemical removal of *Cryptosporidium* in drinking water treatment evaluated 13 published studies on ozone inactivation (Haas et al., 2001). The studies and their respective data were graded (grade A being best on a scale of A – C) based on the methods and description of the methods used to enumerate oocysts, measure infectivity, and the amount of information provided on the disinfection procedures. Out of the 13 studies that were evaluated, only 23 out of 337 data-points (6.8%) were considered A-quality data. A plot of CT (using the initial ozone concentration as C, mg/L) against inactivation for this A-quality data (not accounting for variation in pH or temperature) demonstrated that an average 2-log_{10} inactivation was achieved with a CT of approximately 11 mg.min/L. The B and C rated data were generated using inadequate methods or poorly described and defined procedures and generally displayed much lower levels of inactivation.

All of the A rated data were generated using an animal infectivity assay to measure inactivation whereas much of the B and C rated data were derived from the use of in-vitro excystation procedures. In the absence of routine human feeding trials, infectivity in animal models, particularly mice, has long been considered the "gold standard" for measuring infectivity and disinfection of *C. parvum.* Despite reasonably good agreement between methods, Korich et al. (1990) cautioned that viability estimates based on excystation alone are suspect due to a number of variables affecting the accuracy of the method. Oocysts exposed to 1 mg/L of ozone for 5 minutes at 25°C were still infectious although 100% inactivation was obtained according to the in-vitro excystation method employed. The opposite finding was reported for a variety of viability assays when compared to animal infectivity. Staining oocysts with a range of vital dyes (SYTO-59, SYTO-9, DAPI/PI) demonstrated minimal or marginal reductions in viability following exposure to ozone (Bukhari et al., 2000). Using in-vitro excystation, the average reduction in viability was approximately 0.26-log_{10} following exposure to 0.3 – 0.4 mg/L of ozone for 2 min compared to an average of 0.75-log_{10} when measured by a CD-1 mouse infectivity assay. However, the results of at least one study indicated that a modified in-vitro excystation protocol generated ozone inactivation data equivalent to a mouse infectivity assay and could therefore be used as a substitute for animal assays for measuring ozone inactivation of *C. parvum* (Hirata et al., 2001). Rennecker et al. (1999) reported that ozone inactivation results obtained with a modified in-vitro excystation method based on excysted sporozoite counts were consistent with published inactivation data generated with animal infectivity assays. Proposed CT values for 2-log_{10} inactivation of the Iowa isolate were 9.6, 17.5, and 32.5 mg.min/L at 15°C, 10°C, and 5°C, respectively (Rennecker et al., 1999). In-vitro excystation was also used to demonstrate that ozone inactivation of *C. parvum* (Iowa isolate) was not dependent on pH in the range pH 6 – 10 (Rennecker et al., 2001). Additionally, although the use of epifluorescence microscopy was seen to either under- or overestimate ozone inactivation of *C. parvum* compared to mouse infectivity, depending on the amount of ozonation, flow cytometric analysis of fluorescently stained oocysts was reported to accurately predict inactivation (Gyürék et al., 2001; median oocyst fluorescence was inversely proportional to oocyst infectivity).

Clearly, there is still considerable room for standardization of methods used to measure infectivity and assess the efficacy of disinfectants for *Cryptosporidium.* Following a thorough reanalysis of a selected set of published data, a CT equation based on first order kinetics was proposed to provide guidance to the water industry for the application of ozone for inactivation of *C. parvum* (Clark et al., 2002). Datasets from two studies (Li et al., 2000; Rennecker et al., 1999) was used to develop the model and the authors used the simple Chick-Watson model to describe the inactivation data:

$$Ln(N_t / N_0) = -KCt \quad (1)$$

where K = reaction coefficient (min^{-1})
C = concentration of disinfectant residual (mg/L)
t = period of exposure to the disinfectant (minutes)
N_t = number of viable organisms at time t
N_0 = number of viable organisms at the start of the test (time zero).

The CT equation developed from this study for a given level of inactivation ($Ln(N_t/N_0)$) and temperature (T) was:

$$Ct = 8.5(1.1)^{-T} \times (-Ln(N_t / N_0)) \quad (2)$$

The CT requirements necessary to achieve 1 – 3-log_{10} inactivation of *C. parvum* oocysts based on this equation are given in Table 1.4. The authors also provided an approach for calculating a safety factor; incorporation of safety factors into disinfection requirements is considered a necessary approach to account for the variability in the underlying data that is used to develop disinfection models. Adopting the 90% confidence level as an appropriate safety factor led to an approximate doubling of the CT required to achieve 2-log_{10} inactivation (Clark et al., 2002).

A further factor that needs to be considered in the design and interpretation of disinfection experiments is the concentration of target microorganisms used to spike reactors. A study with *Giardia muris* demonstrated that it was more difficult to achieve inactivation (longer exposure time at a given ozone concentration) of cysts when the initial density was low (5,000 cysts/mL) than when it was high (50,000 cysts/mL; Haas and Kaymak, 2003). Possible explanations for this phenomenon are: higher densities of microorganisms result in the formation of more disinfection byproducts, some of which may have inactivation properties; the cysts had quorum sensing function(s) which altered the response of a population of cysts at higher densities. Quorum sensing is a widely recognized property in bacteria whereby the accumulation of signaling molecules enables a bacterium to sense the overall cell density in a population and modify its response to stimuli and stresses.

Because of its instability in water, ozone decays rapidly and so will not maintain a residual disinfection activity in the distribution system. Consequently, ozone will be applied in conjunction with a relatively low concentration of a second disinfectant, typically chlorine or chloramines, which will maintain a residual. Such combinations of mixed oxidants generally result in synergistic effects. At 22°C in 0.05 M phosphate buffer, ozone pre-treatment increased the efficiency of free chlorine for inactivation of *C. parvum* (Iowa isolate) 4 – 6 times (Li et al., 2001) as measured by infectivity in CD-1 mice. Synergistic inactivation was also reported for ozone pre-treatment followed by free chlorine or monochloramine with the synergy increasing at lower temperatures (Rennecker et al., 2000). The secondary inactivation rate was up to 2.8 (free chlorine) and 9.2 (monochloramine) fold faster than the respective primary inactivation rates at 10°C.

Table 1.4
Ozone CT requirements based on equation 2* for inactivation of *C. parvum*

	CT required to achieve indicated inactivation:		
Temperature	1-$\log_{10}$	2-$\log_{10}$	3-$\log_{10}$
1°C	17.8	35.6	53.4
5°C	12.1	24.3	36.5
15°C	4.7	9.4	14.1

* from Clark et al., 2002.

Table 1.5 provides a summary of selected ozone disinfection studies for *C. parvum*. Since all of the studies up to 1998 were reviewed by Haas et al. (2001), only more recent studies were included in the present review. In addition, studies were selected to demonstrate the range of procedures employed and no attempt was made to grade the quality of the data.

Table 1.5
Summary of selected ozone inactivation studies for *Cryptosporidium parvum*

Reference	Reactor (volume or flow rate)	Infectivity/ viability model	Matrix	Temp. (°C)	Infection/viability detection	Inactivation calculation	Isolate* and host	Oocyst age	Inactivation†
Bukhari et al., 2000	Batch (80 mL)	CD-1 mice	Deionized water	22	Microscopy on stained ileum sections	Logistic dose response	Iowa Calf	14 – 84 days	2.2-$\log_{10}$ at CT = 0.6
Driedger et al., 2001	Semi-batch (200 mL)	In-vitro excystation	0.01 M PO_4 buffer	1 – 20	Microscopy	No. of excysted sporozoites	Iowa Calf	27 – 90 days	k = 0.15 – 1.65‡
Gyürék et al., 2001	Batch (500 mL)	SYTO-9 staining	0.05 M PO_4 buffer	1 – 22	Flow cytometry	Increased fluorescence	Iowa Calf	≤ 91 days	2-$\log_{10}$ at CT = 40, 5°C
Hirata et al., 2001	Semi-batch (500 mL)	SCID mice	10 mM PBS	3 – 30	Fluorescence microscopy on fecal samples	MPN, 5-fold dilutions	HNJ-1 SCID mice	Not provided	2-$\log_{10}$ at CT = 12, 10°C
Li et al., 2000	Batch (200 mL)	CD-1 mice	0.05 M PO_4 buffer	1 – 22	Flow cytometry on homogenates	Logistic dose response	Iowa Calf	≤ 90 days	1.5-$\log_{10}$ at CT = 40, 1°C
Oppenheimer et al., 2000	Single CSTR (430 mL)	CD-1 mice	Natural waters	3 – 23	DIC microscopy on fecal samples	MPN, 3 dilutions, 4 mice per dilution	Iowa Calf	≤ 56 days	1-$\log_{10}$ at CT = 3.8 – 9, 10°C
Owens et al., 2000	Pilot-scale (6.4 L/min)	BALB/c mice	Filter effluent	24.5	Fluorescence microscopy on homogenates	Logistic dose response	Iowa Calf	≤ 60 days	2-$\log_{10}$ at CT = 6 k = 0.86‡
Owens et al., 2000	Pilot-scale (6.4 L/min)	In-vitro excystation	Filter effluent	23.6	Phase contrast microscopy	Ratio of excysted to intact oocysts	*C. muris* RN66 CF-1 mice	≤ 30 days	2-$\log_{10}$ at CT = 8
Rennecker et al., 1999	Semi-batch (200 mL)	In-vitro excystation	0.01 M PO_4 buffer	5 – 30	Phase contrast microscopy	No. of excysted sporozoites	Iowa C57/BL-6 mice	< 90 days	1-$\log_{10}$ at CT = 3.6, 20°C k = 0.80‡, 20°C
Rennecker et al., 2001	Semi-batch (200 mL)	In-vitro excystation	0.01 M PO_4 buffer	20	Phase contrast microscopy	No. of excysted sporozoites	Iowa Calf	35 – 90 days	k = 0.74‡

* All isolates were *Cryptosporidium parvum* unless indicated otherwise.
† CT units = mg.min/L. ‡ Range of inactivation rate constants (L/mg.min) at lowest and highest temperatures tested.

CHAPTER 2
STATEMENT OF PROBLEM

The water industry requires infectivity assays to provide much needed information on the occurrence of infectious *Cryptosporidium* oocysts in raw and finished water and the efficacy of disinfection and treatment strategies.

The three techniques available to determine *C. parvum* infectivity are human volunteers, animal models, and in-vitro cell culture. Human volunteer studies are not practical for use on a routine basis by the water industry. Mouse infectivity assays are currently the accepted technique for measuring *C. parvum* infectivity (traditionally considered as the "gold standard") but the expense and labor intensive nature of mouse infectivity assays limits their practicality for routine monitoring. Also, some strains/isolates of *C. parvum* do not efficiently infect mice or other animal models. Consequently, in-vitro cell culture provides a promising and practical alternative technique for determining infectivity of waterborne *Cryptosporidium*. Cell culture techniques are routinely used for measuring the infectivity of a wide range of human pathogens and a variety of different cell culture systems and detection methodologies have been applied to *C. parvum*. However, the water industry does not yet recognize cell culture as an acceptable technique because the relationship between cell culture-based infectivity and animal infectivity has not been thoroughly investigated. Although many cell culture methods have been developed for *C. parvum*, it has not been proven that infectivity in cell culture is a good indicator of the ability of oocysts to cause infections in animals. Also, the apparent efficacy of disinfectants may be affected by the method used to measure inactivation; cell culture-based procedures may not give the same results as animal infectivity assays. In addition, different cell lines used for in-vitro cell culture demonstrate varying susceptibility to *C. parvum* infection. Therefore, studies are needed to assess the correlation between cell culture and animal-based infectivity of *C. parvum* to determine whether the different methods vary in their sensitivity to infection, dose responses, or ability to demonstrate oocyst inactivation.

A key concern in producing the amount of *Cryptosporidium* disinfection data which is needed for regulatory development or for testing disinfectant efficacy is the lack of a secondary reference method for *Cryptosporidium* inactivation. A study designed to determine which of four in-vitro viability surrogates could serve to replace animal infectivity demonstrated that none of the in-vitro surrogates were reliable indicators of infectivity (Bukhari et al., 2000). The results of this study indicated that for ozone and UV light, animal infectivity was the only reliable indicator for oocyst infectivity. This severely limits the amount of disinfection work that can be done, as animal infectivity studies require specialized facilities found in only a limited number of laboratories and are extremely labor intensive and expensive. Consequently, alternative infectivity methods, such as cell culture, need to be adopted, but only if they demonstrate equivalency or superiority to animal models.

PROJECT OBJECTIVES

Cell culture-based infectivity assays provide a useful tool that can be used to gain valuable information about *Cryptosporidium* in drinking water. They are generally practical and relatively straightforward techniques that can be used to measure the efficacy of disinfectants, the infectivity of oocysts detected in water, and the survival and persistence of oocysts in the environment. However, further research is necessary to compare cell culture with the "gold standard" of mouse infectivity before the method can be widely adopted by the water industry. Therefore, the overall objective of the research described in this report was to compare in-vitro cell culture-based systems with an animal model for determining the infectivity of *Cryptosporidium parvum* oocysts and to compare the assays for measuring disinfection of oocysts. The experimental aspects of the project were defined by the following specific objectives:

1. Use at least four different isolates of *C. parvum* to compare the dose response and sensitivity of cell culture and mouse infectivity assays.
2. Compare cell culture and animal infectivity assays for determining the efficacy of chemical (ozone) and physical (UV) disinfectants.
3. Perform an experimental assessment of ozone inactivation of *C. parvum*.
4. Compare infectivity methods with oocysts exposed to environmental water samples.
5. Determine the reproducibility and variability of the methods by statistical analysis of all data.

CHAPTER 3
MATERIALS AND METHODS

A schematic outline of the overall research approach is provided in Figure 3.1. In Phase 1 of the project, a single animal infectivity model, neonatal CD-1/ICR mouse, was compared with three cell culture assays involving three cell lines and two methods for detecting infection, for measuring the infectivity of fresh, untreated *C. parvum* oocysts. The CD-1 mouse is a well established model for analyzing *C. parvum* infectivity and disinfection and has been used for inter-laboratory comparisons of mouse assays (Korich et al., 2000a). The average ID_{50} for the Iowa isolate of *C. parvum* in CD-1 mice, based on many dose response experiments was 100 (Korich et al., 2000b). In Phase 2, cell culture and mice were compared for assessing oocyst inactivation by UV light and ozone.

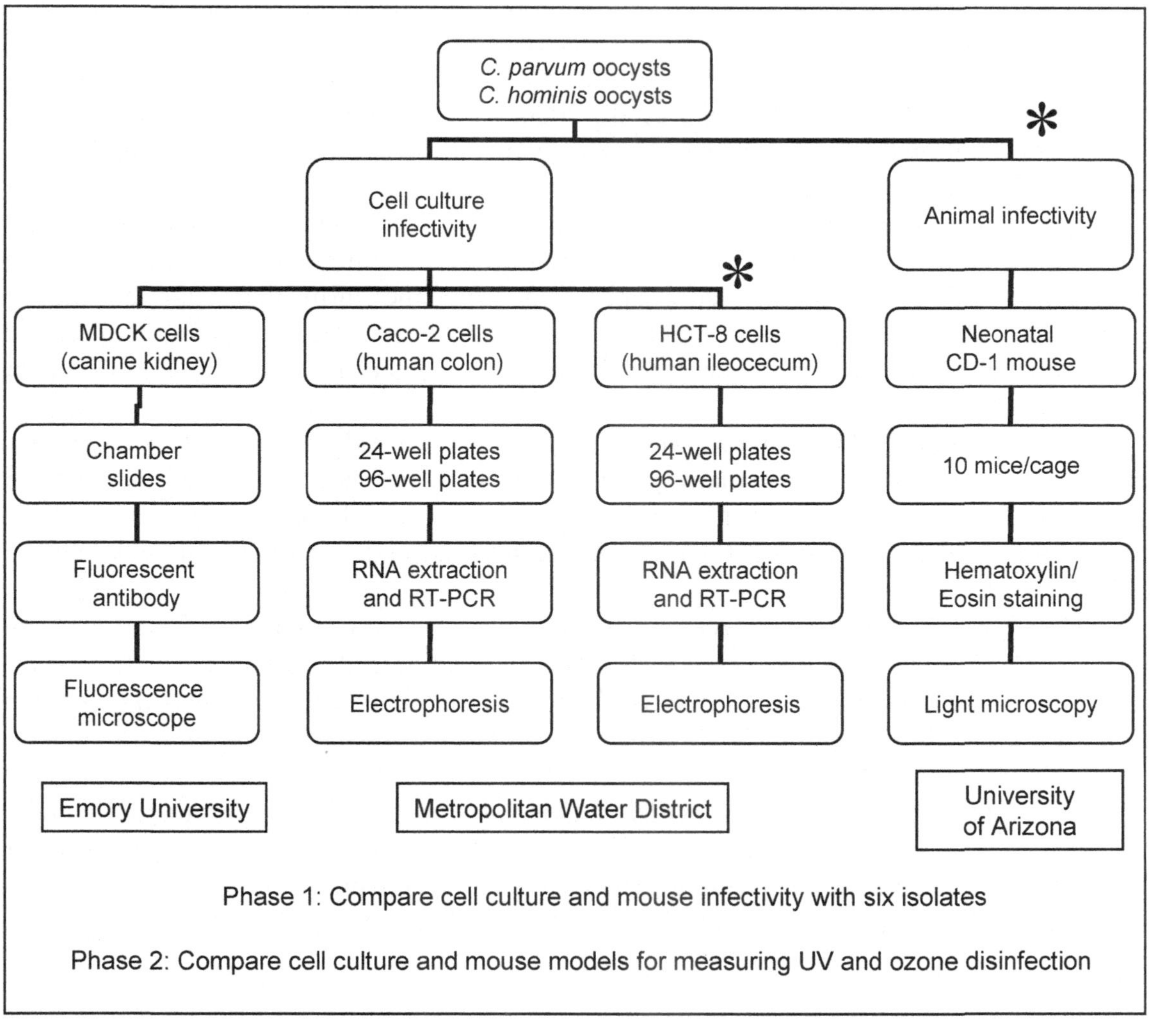

Figure 3.1 Overall project approach. The primary infectivity assays used for comparisons with disinfected oocysts are indicated with an asterisk.

All animals used in this research for propagation of oocysts and infectivity assays were housed, handled, and monitored in an accredited animal care facility at the University of Arizona in accordance with the Animal Welfare Act of 1985, the Public Health Service Policy of 1986, the Good Laboratory Practice Act of 1978, and the National Academy of Sciences Guide for the Care and Use of Laboratory Animals (1996).

ISOLATES, OOCYST PRODUCTION AND PURIFICATION

The isolates used in this investigation are described in Table 3.1. Five of them are categorized as zoonotic or genotype 2 isolates of *C. parvum* while one was genotype 1. Under the recently proposed division of genotype 1 and genotype 2 *C. parvum* into 2 separate species (Morgan-Ryan et al., 2002), throughout this report, isolates previously called genotype 1 *C. parvum* are referred to as *C. hominis* while genotype 2 isolates retain the name *C. parvum*.

The Iowa isolate of *C. parvum* has been well characterized and has been used extensively in experimental infections involving cell culture, animal models, human feeding trials, disinfection, and genetic fingerprinting studies (Bukhari et al., 1999; Chappell et al., 1996; Dupont et al., 1995; Finch et al., 1993; Korich et al., 1990; Korich et al., 2000a; Mofidi et al., 2001; Okhuysen et al., 1999; Rochelle et al., 2000a). The isolate was originally obtained from Harley Moon (National Animal Disease Center, Ames, Iowa) and is maintained and propagated on a routine basis at the University of Arizona.

The Maine isolate was responsible for an outbreak of cryptosporidiosis traced to apple cider (Millard et al., 1994). Oocysts were isolated directly from apple cider, the press used for preparing the cider, and a calf stool specimen from the farm that supplied the apples. Experimental infection studies showed that this isolate was infective for both mice and calf hosts; it was characterized as a genotype 2 isolate using the TRAP-C2 polymorphic marker (Peng et al., 1997).

The Moredun isolate was originally isolated from a red deer calf (*Cervus elaphus*) in 1986 but has been passaged for many years in sheep. It has been classified as a genotype 2 isolate using a variety of polymorphic markers (Spano et al., 1997; Carraway *et al.*, 1997) but it is not identical to the Iowa isolate. It has also been used in human feeding trials (Okhuysen et al., 2002).

The TAMU isolate was recovered from humans who became infected handling an infected foal. It has subsequently been used in human feeding trials (Okhuysen et al., 1999).

The Glasgow isolate was responsible for an outbreak of cryptosporidiosis linked to contaminated drinking water (H. V. Smith, pers. comm). It has not been studied extensively in experimental infections.

The isolate of *C. hominis*, HIV1487, was isolated from a human immunodeficiency virus (HIV) patient attending a clinic in Lima, Peru (R. Gilman, pers. comm). Isolates used in this investigation were characterized with at least two genetic markers (β-tubulin and COWP genes) to determine their genotype as described previously (Rochelle et al., 2001a).

Table 3.1
***Cryptosporidium* isolates used in this study**

Isolate	Original host	Species (genotype)*	Source	Reference
Glasgow	Human	*C. parvum* (genotype 2)	H. V. Smith, Scottish Parasite Diagnostic Laboratory, Glasgow, Scotland	H. V. Smith, pers. comm.
Iowa	Cow	*C. parvum* (genotype 2)	H. Moon, National Animal Disease Center, Ames, Iowa	Korich et al., 1990
Maine	Human	*C. parvum* (genotype 2)	M. J. Arrowood, Centers for Disease Control, Atlanta, Georgia	Millard et al., 1994
Moredun	Deer	*C. parvum* (genotype 2)	S. E. Wright, Moredun Research Institute, Peniculk, Scotland	Blewett et al., 1993
TAMU	Horse	*C. parvum* (genotype 2)	C. L. Chappell, University of Texas Health Science Center, Houston, Texas	Okhuysen et al., 1999
HIV1487	Human	*C. hominis* (genotype 1)	R. Gilman, Johns Hopkins University, Baltimore, Maryland	R. Gilman, pers. comm.

* Species names are based on the recently proposed renaming of *C. parvum* genotypes (Morgan-Ryan et al., 2002). The previously used nomenclature for *C. parvum* genotypes is included in parentheses.

Oocysts of *C. parvum* were propagated by artificial infection of Holstein calves housed in the University of Arizona Central Animal Facility (accredited by the American Association of Accreditation of Laboratory Animal Care). Calves were fed 2×10^8 oocysts suspended in sterile water within 4 to 12 h of birth, followed 1 - 2 h later by 1.9 L of colostrum. The calves received prophylactic doses of oral Rota/Corona Virus and *Escherichia coli* vaccines. Subsequently, the calves were maintained on a diet providing 2200 kcal/day supplied by a combination of milk replacer, non-fat milk powder, and electrolyte mix. All procedures for calf care and infection were approved by the University of Arizona's Institutional Animal Care and Use Committee. The total daily fecal output from infected calves was collected, screened through graded sieves, and concentrated by centrifugation. Oocysts were purified from fecal material by sequential density gradient centrifugation through sucrose and cesium chloride gradients as described previously (Arrowood and Donaldson, 1996). Purified oocysts were suspended in phosphate-buffered saline (PBS) containing 0.01% Tween 20, 100 U/mL of penicillin, 100 μg/mL streptomycin, and 100 μg/mL of gentamicin, and stored at 4°C until they were used for infection. The excystation rate was determined for each batch of freshly purified oocysts. Most experiments were performed with freshly purified oocysts within 10 – 18 days of shedding. Rigorous quality control procedures ensured that each infection trial used oocysts produced by the same calf on the same shedding day. Lot numbers were assigned to each calf on the day of infection and batch numbers were assigned to the oocysts on the day they were shed. These lot-batch numbers were used to track the oocysts from the production laboratory, through the flow cytometry facility, and to each of the three laboratories for assessment of infectivity.

Oocysts of *C. hominis* were obtained from a human HIV patient. Since these oocysts could not be propagated in calves, they were purified and inoculated directly onto cell monolayers 62 days after isolation from the infected individual.

For negative infection controls, oocysts were inactivated by heating at 70°C for 30 min. For transport between the various laboratories involved in this investigation, oocysts were packed on ice, packaged and labeled as required by the International Air Transport Association Dangerous Goods Regulations and shipped via Federal Express. Chain of custody forms accompanied all shipments of oocysts.

Beginning in June 2001, a quality assurance program was initiated for all lots of oocysts received at MWD. Immediately upon receipt, a portion of each oocyst lot was examined microscopically to assess the amount of general debris and contaminating microorganisms. Aliquots were also inoculated onto nutrient agar and m-endo agar plates (incubated at 37°C for 24 h), Sabaroud-dextrose agar plates (incubated at 28°C for 7 d), and cell culture growth medium (including antibiotics) to assess the level of bacterial and fungal contamination. Experience has demonstrated that *C. parvum* oocyst preparations are sometimes contaminated with *Escherichia coli* (Table 3.2). Some of these contaminating strains of *E. coli* were resistant to the antibiotics used in the cell culture medium. Any putative positive coliforms on the m-endo plate were identified using API20E strips following the manufacturer's recommendations (Biomerieux, Hazelwood, MO). The presence of antibiotic resistant bacteria in oocyst preparations highlights the importance of a routine QC procedure for oocysts, particularly when they are intended for use in cell culture-based assays.

In-Vitro Excystation of *C. parvum* Oocysts

The in-vitro excystation efficiency was determined for most preparations of oocysts used in this investigation (Table 3.2). Oocysts were incubated in acidified Hank's balanced salt solution for 1 h at 37°C followed by 1 h at 37°C in a solution containing 0.8% trypsin and 0.75% sodium taurocholate, and finally 30 min at room temperature in the latter solution. The rate of excystation was determined by microscopic examination to enumerate the number of excysted and unexcysted oocysts.

Table 3.2
Quality control of oocysts used for infectivity and disinfection experiments

Isolate	Lot number	Excystation rate (%)	Age (days)*	Contamination in oocyst stock†			
				NB‡	m-endo§	*E.coli***	RPMI + Abs††
Glasgow	01127G-04	98	8 – 14	nd	nd	nd	nd
Iowa	00104-11	95.1	15	nd	nd	nd	nd
Iowa	10327-01	98.1	15	nd	nd	nd	nd
Iowa	10423-28	96	16	nd	nd	nd	nd
Iowa	10514-18	99	16	nd	nd	nd	nd
Iowa	10618-23	98.8	23	+	+	+	–
Iowa	10806-13	98.8	14 – 27	+	–	–	–
Iowa	10910-16	95	29	–	–	–	–
Iowa	11008-16	97	13	–	–	–	–
Iowa	11105-11	93.6	15	+	–	–	–
Iowa	20102-10	98.4	12	+	+/+‡‡	+/+‡‡	+/–‡‡
Iowa	20204-11	97.7	8 – 31	+	–	–	–
Iowa	20225-06	94.7	12	+	+	–	–
Iowa	20311-17	98	21	+	–	–	–
Iowa	20422-27	97.2	10	–	–	–	–
Iowa	20506-14	96.7	26	–	–	–	–
Iowa	20603-10	94	14	–	–	–	–
Iowa	20617-25	98	13	+	+	+	–
Iowa	20708-17	96.6	25	–	–	–	–
Iowa	20722-29	97.5	27	–	–	–	–
Iowa	21105-13	92	29	–	–	–	–
Iowa	30107-14	95.8	11	–	–	–	–
Iowa	30310-18	93.4	26	–	–	–	–
Maine	00905AC-13	97.2	14 – 20	nd	nd	nd	nd
Moredun	00228M-06	98	11 – 37	nd	nd	nd	nd
Moredun	10102M-09	93	25	nd	nd	nd	nd
TAMU	00606T-09	94.6	15 – 50	nd	nd	nd	nd
TAMU	10212T-19	98	15 – 19	nd	nd	nd	nd

* Number of days post-shedding from propagation animal at the time of infection or disinfection.

† Quality control program was instituted part way through this project so some oocyst lots were not screened for bacterial contamination (nd).

‡ Nutrient broth.

§ Growth on m-endo agar (selective for coliforms).

** Identification of *E. coli* as primary contaminant, based on API20E strips.

†† Growth in RPMI cell culture media containing antibiotics.

‡‡ Oocysts were screened for contamination immediately upon receipt and following a second cesium chloride gradient purification. Results are reported as initial screen/second screen.

Genotyping Isolates

Isolates were genotyped by PCR targeting the β-tubulin and *Cryptosporidium* oocyst wall (COWP) protein genes using previously described methods (Rochelle et al., 1999b; Rochelle et al., 2001a). Briefly, DNA was extracted from purified oocyst preparations and amplified using the primers described in Table 3.3, Tth thermostable polymerase (Roche Molecular Biochemicals, Indianapolis, IN), 1.5 mM $MgCl_2$, and hot start conditions with a 60°C annealing temperature. Amplification products were purified by passage through Wizard columns (Promega, Madison, WI) and cloned into the PCR2.1-TOPO vector by TA cloning (Invitrogen, Carlsbad, CA) and transformed into *Escherichia coli* TOP10 cells using standard procedures (Sambrook et al., 2001). Plasmid DNA was purified from 25 mL broth cultures using Qiagen columns (Qiagen, Santa Clarita, CA) and cloned insert DNA was sequenced in both directions (Integrated Biotechnology Laboratories, University of Georgia, Athens, GA).

The Iowa, Moredun, TAMU, Maine, and Glasgow isolates were all confirmed as *C. parvum* (previously genotype 2 *C. parvum*). Out of eight human isolates recovered from HIV patients in Lima, Peru, three were *C. hominis* (previously genotype 1 *C. parvum*), three were *C. parvum* (previously genotype 2), and two did not produce amplification products wither set of primers. These isolates were not collected specifically for this project and were stored for up to 6 months in potassium dichromate prior to DNA extraction and amplification. Also, no indication was provided regarding the severity of infection or the number of oocysts in the samples. Examples of β-tubulin gene sequences used for genotyping isolates are provided in Figure 3.2. Previous genotyping has demonstrated considerable intra- and inter-isolate heterogeneity within the β-tubulin of *C. parvum* (Rochelle et al., 2001a), so the heterogeneity within the Maine isolate before and after calf-propagation and the differences between independent clones of the TAMU isolate was not unexpected. Typically, this heterogeneity is only observed within the poly-thymine regions and does not affect the nucleotide positions that are diagnostic for the *C. parvum–C. hominis* differentiation. Heterogeneity within the poly-thymine regions may be genuine sequence differences or artifacts of the PCR, cloning, and sequencing procedures. Simple homopolymeric repeats such as the poly-thymine regions are particularly prone to PCR introduced errors caused by "slippage" of the polymerase and compression artifacts (Hayashi, 1994).

Enumeration of Oocyst Challenge Doses

Accurate enumeration of oocyst challenge doses is a critical component of measuring oocyst infectivity and inactivation. This ensures that differences in the amount of infectivity or inactivation are real differences and not apparent differences resulting from inaccurate or imprecise enumeration of oocyst doses. A consistent problem with *C. parvum* dose response infectivity and disinfection studies is determining the number of oocysts which are actually used. Frequently, oocysts are enumerated by hemacytometer counting or fluorescence microscopy following staining with fluorescently labeled antibodies. Based on these counts, dilutions are then made which are used for infectivity or disinfection. Such enumeration methods can greatly under- or overestimate the number of oocysts present, due to counting errors, oocyst clumping, and dilution errors. To overcome this potential source of error, in phase 1 of the project, which involved comparison of mouse and cell culture infectivity assays with untreated oocysts, the purified oocysts were enumerated and sorted into challenge doses by flow cytometry using an Epics Elite flow cytometer (Coulter Corporation, Miami, FL) at the Wisconsin State Laboratory of Hygiene (Madison, WI). This ensured the validity of the dose-response curve data by providing the same standardized oocyst preparations for each of the three laboratories and each infectivity model. Cytometers were calibrated using fluorescent polystyrene beads of uniform size. Following initial calibration, a preparation of fluorescently labeled oocysts (Crypt-a-Glo fluorescent antibody, Waterborne, Inc., New Orleans, LA) drawn from each oocyst lot was sorted onto ten well slides and observed under epifluorescence optics at 200× magnification. Target levels from the population of unlabeled oocysts were then sorted into 50 mL polypropylene tubes containing 0.01% Tween-20 in sterile water. Ongoing precision recoveries (OPRs) consisting of well slide enumerations of fluorescently labeled oocysts were performed every tenth sort and following the final sort. Replicate samples containing precisely defined numbers of oocysts were prepared and shipped to each of the three analysis laboratories for dose response experiments. A trip control was included for each lot of oocysts shipped in order to monitor any deleterious effects on oocysts associated with handling/storage during shipping. The average coefficient of variation, based on 13 microscopic counts of flow cytometry sorted oocysts for each isolate, was 2.1% (range of mean counts was 95.4 – 101.5).

Genotype 2, *Cryptosporidium parvum*	
IOWA	GACTTTTTTTTTTTCTTTTTTTTTCTTTGAAA
MAINE, before propagation	GACTTTTTTTTTTTCTTTTTTTTTCTTTGAAA
MAINE, after propagation	GACTTTTTTTTTT-CTTTTTTTTTCTTTGAAA
TAMU, before propagation	GACTTTTTTTTTTTCTTTTTTTTTCTTTGAAA
TAMU, after propagation[1]	GACTTTTTTTTTTTCTTTTTTTTTCTTTGAAA
TAMU, after propagation[2]	GACTTTTTTTT---CTTTTTTTTTCTTTGAAA
Human[a]	GACTTTTTTTTTTTCTTTTTTTT-CTTTGAAA
Human[b]	GACTTTTTTTTTTTCTTTTTTTTTCTTTGAAA
Genotype 1, *Cryptosporidium hominis*	
Human[c]	GACTCTTTTTTTTTCTTTTTTT--CTCTGAAA
Human[d], HIV1487	GACTCTTTTTTTTTCTTTTTTT--CTCTGAAA
Other isolates	
C. meleagridis (AF323575)	GATTCCTTTTTTTTTTTCTTT---CTTCGAAA
Muskrat isolate (AF303056)	AACTTTTCTTTTTTTTTTCTTTTTGTTTCTAT

Figure 3.2. Alignment of nucleotides in a genotype-diagnostic region of the β-tubulin gene of *Cryptosporidium* spp. All sequences were generated for this project except the *C. meleagridis* and muskrat isolates (referenced by GenBank accession number) which were included as examples of more distantly related species. The Maine and TAMU isolates were analyzed before and after propagation at UA in calves. For the TAMU isolate, sequence fragments of two individual clones from a single oocyst preparation are shown ([1] and [2]). The sequences of four human isolates are also displayed ([a] – [d]). Diagnostic nucleotide polymorphisms are boxed (T, *C. parvum*; C, *C. hominis*).

Oocyst Enumeration for Disinfection Experiments

For UV and ozone disinfection studies, oocyst stock preparations were enumerated using a modified Neubauer hemacytometer to generate a mean of at least five replicate counts. Oocysts were spiked into disinfection reactors to achieve particular concentrations based on these enumerations. Following exposure to disinfectant, oocysts were diluted, stained with a fluorescein isothiocyanate labeled monoclonal antibody, and enumerated by epifluorescence microscopy. At least five 100 μL aliquots of diluted oocysts were placed on two-well Superstick slides (Waterborne, Inc., New Orleans, La.) or three-well slides (Meridian Diagnostics, Cincinnati, OH) and allowed to dry overnight at room temperature. Immunofluorescence counts were performed on 100 μL aliquots of each challenge dose because this was the same volume that was used to inoculate cell monolayers. Diluted Aqua-Glo GC Direct antibody (Waterborne, Inc.) was applied and incubated at room temperature for 30 min. Following rinsing in PBS, slides were mounted (Merifluor mounting medium; Meridian Diagnostics) and examined using a Nikon Optiphot fluorescence microscope with excitation and emission wavelength filters of 450 – 490 nm and 515 nm, respectively. Each challenge dose was calculated as the mean of at least five and up to 10 individual counts. Enumeration of oocyst challenge doses during this project resulted in a generally acceptable level of precision (Figures 3.3 and 3.4).

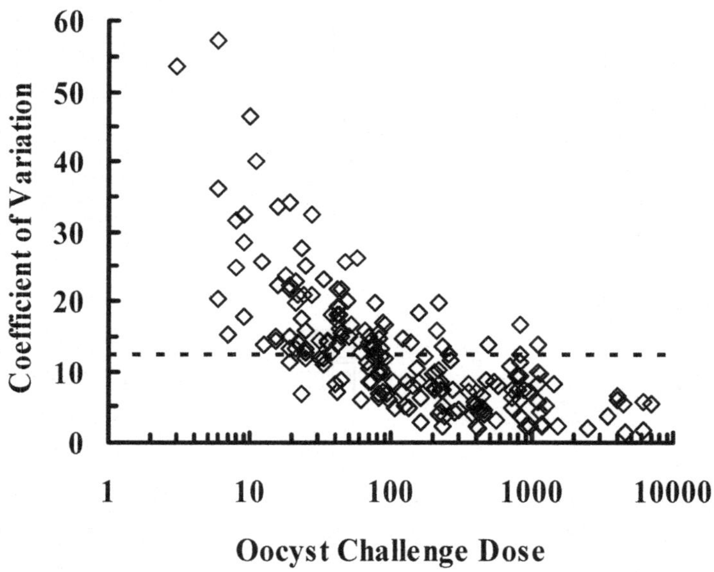

Figure 3.3 Relationship between oocyst challenge dose for disinfection studies and the coefficient of variation (CV) for enumeration of oocysts by immunofluorescence (N = 204 challenge doses with a minimum of 5 enumerated wells per dose). The mean CV (12.5%) is represented by the horizontal dashed line.

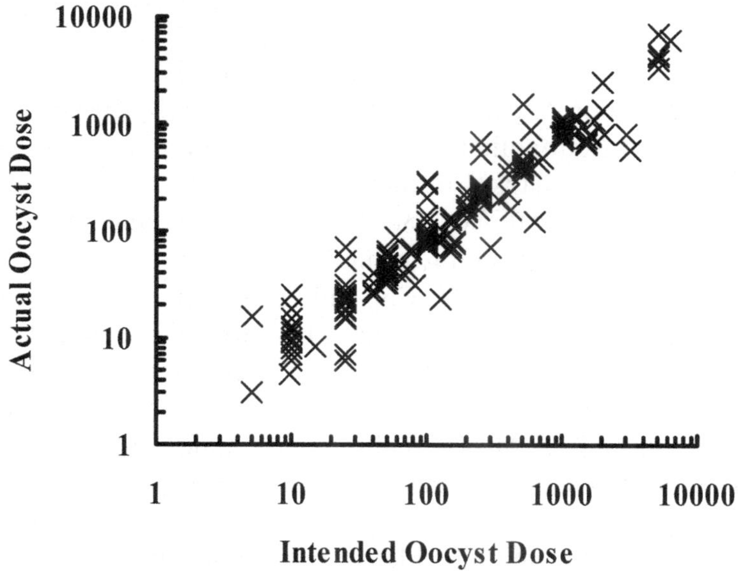

Figure 3.4 Correlation between the intended oocyst challenge dose (based on hemacytometer enumeration) and the actual oocyst dose, based on immunofluorescence enumeration for disinfection studies.

The mean coefficient of variation (CV) for all of the oocyst challenge doses following disinfection and used to infect HCT-8 cell cultures was 12.5% with a range of 1.5 – 57.3%. Figure 3.4 demonstrates that the actual oocyst dose inoculated onto cell monolayers correlated strongly with the intended dose based on hemacytometer enumeration. As would be expected, there was an inverse relationship between the enumerated oocyst doses and CV.

MOUSE INFECTIVITY ASSAYS

Dams and neonatal CD-1/ICR mice (5- to 7- days old) were obtained from Charles River Laboratories, Inc. (Wilmington, MA) and neonates were inoculated by delivering a 10 μl aliquot of each dose of oocysts (doses ranged from 100 to 50,000 oocysts), suspended in sterile water, to the back of the throat using a calibrated P-10 pipette (Rainin Instruments, Oakland, CA). Pipets were calibrated every six months by Continental Lab Services (San Diego, CA) and calibration reports are archived at the Sterling Parasitology Laboratories (University of Arizona). Cages were labeled with date of infection, oocyst lot number, challenge doses, number of pups inoculated and study code. Seven days after inoculation, formalin fixed, slide mounted sections of the terminal ileum (5 μm × 2-3 cm, removed at necropsy) were stained with hematoxylin and eosin (Korich et al., 2000a) and infections were detected by microscopic examination. The ileum was chosen because it is the first region of the gut to be colonized by the parasite and harbors the majority of development stages (Current and Reese, 1986). Evidence of infection was defined as the observation of *C. parvum* parasite developmental stages in the microvilli of any of the prepared histological sections. Tissue blocks and labeled slides are archived for 5 years.

CELL CULTURE PROCEDURES FOR INFECTIVITY STUDIES

Monolayers of the human ileocecal adenocarcinoma cell line HCT-8 (ATCC CCL-244; American Type Culture Collection, Rockville, MD) were initially grown and maintained in 75 cm^2 and 150 cm^2 flasks at Metropolitan Water District. For infectivity assays, cells were grown in 24-well or 96-well culture plates (Corning Costar, Corning, NY). Cells were cultured in RPMI-1640 medium (Sigma Chemical Co., St. Louis, MO) supplemented with 4 mM L-glutamine, 30 mM HEPES, pH 7.3, 100 U/mL penicillin, 100 µg/mL each of streptomycin and kanamycin, and 0.25 µg/mL amphotericin B. Cell culture growth medium was supplemented with 10% fetal bovine serum (FBS; HyClone, Logan, UT) for maintenance of uninfected cells and 2% FBS for cells following inoculation with oocysts. Cells were incubated in a humidified incubator at 36°C in an atmosphere containing 5% CO_2. Caco-2 cells (ATCC HTB 37) were grown under the same conditions as HCT-8 cells using Eagle's minimal essential medium (Sigma) supplemented with the additives described above, 15% FBS for maintenance of uninfected cells, and 5% FBS for growth following inoculation with oocysts. Oocysts were resuspended in cell culture media, and inoculated onto monolayers that were 80 – 100% confluent and had a uniform microscopic appearance. Inoculated monolayers were incubated for 48 h at 36°C in 5% CO_2. Infectivity was not measured in cell lines that had been passaged more than 30 times.

Madin-Darby canine kidney (MDCK) cells (ATCC CCL 34) were grown at the Atlanta VA Medical Center in serum-free Ultraculture medium (Biowhitaker, Walkersville, MD), supplemented with penicillin, streptomycin, and L-glutamine as above in 2-well culture chamber slides (Nunc, Naperville, IL) and incubated at 37°C in 5% CO_2. Oocysts were resuspended in cell culture media and inoculated onto confluent monolayers and incubated at 37°C for 48 h.

RT-PCR Detection of Infection in Cell Cultures

Following incubation of inoculated monolayers for 48 h, infection in HCT-8 and Caco-2 cells was detected by reverse transcriptase-PCR (RT-PCR) amplifying a 361 bp fragment of *C. parvum*-specific heat shock protein (hsp70) mRNA using a modification of a previously described procedure (Rochelle et al., 1997). This procedure made no attempt to use RT-PCR detection in a quantitative manner, but simply as a measure of presence or absence of infection in each of the multiple cell monolayers that were infected with each challenge dose of oocysts. The approach is analogous to the standard procedure used to quantify infectivity in mice in which infectivity is expressed as the proportion of animals that become infected.

All glassware used for RNA extraction reagents was baked at 180°C overnight, and all reagents treated with 1% diethyl pyrocarbonate (DEPC) or made with DEPC-treated sterile distilled water to destroy contaminating RNA degrading enzymes. DEPC was removed from water and reagents by autoclaving after incubation at room temperature for 8 h. Total RNA was extracted using a S.N.A.P. Total RNA isolation kit (Invitrogen, Carlsbad, CA), which incorporates a DNase digestion step to eliminate DNA carryover, or RNeasy Kits (Qiagen, Valencia, CA) following the manufacturer's recommendations. Digestion of extracted RNA with DNase (Ambion, Austin, TX) was included in the procedures to ensure that the final RT-PCR signal resulted from amplification of mRNA derived from sites of active infection and not from carryover of DNA that may be co-extracted from inactive oocysts remaining on the cell monolayers. The final RNA pellet was resuspended in 100 μL of DEPC-treated sterile deionized water. Logistical reasons sometimes prevented the immediate extraction of RNA from cell monolayers following incubation so cell monolayers were preserved in RNA Later (Ambion) and stored at -20°C. In-house comparisons demonstrated no difference in RT-PCR signals and infectivity levels resulting from immediate RNA extraction or extraction from RNA Later-preserved cells. The reverse transcription (RT) and polymerase chain reactions (PCR) were performed separately in all cases. RT reactions were performed using an oligo $(dT)_{16}$ primer, MuLV reverse transcriptase, 1 U of RNase inhibitor per reaction (all supplied by Applied Biosystems, Foster City, CA), 200 μM each of dATP, dCTP, dGTP, and dTTP (Roche Molecular Biochemicals), and 10 μL of extracted RNA in PCR buffer II (Applied Biosystems) containing 5 mM $MgCl_2$ in a 20-μl reaction volume. Incubation conditions for the RT reactions were 10 min at 25°C, 60 min at 42°C, and 5 min at 99°C. Amplification reactions were mediated by 0.025 U/μL of Platinum Taq DNA polymerase (Invitrogen) and 0.25 μM of the *C. parvum*-specific primers cphsp2423F and cphsp2764R (Table 3.3) with an annealing temperature of 53°C. These primers are broadly applicable to RT-PCR based infection detection assays because they are specific for an array of *C. parvum* and *C. hominis* isolates that are potentially pathogenic for humans (Figure 3.5). Amplification reactions contained 10 mM Tris-HCl (pH 8.3), 50 mM KCl, 1.5 mM $MgCl_2$, 200 μM each of dATP, dCTP, dGTP, and dUTP, 0.01 U/μL of uracil DNA glycosylase (UDG; Roche Molecular Biochemicals), and 10 μL of the cDNA produced by the RT reaction. UDG and dUTP instead of dTTP were included as a quality control measure to prevent carryover contamination of PCR products from earlier reactions. Reactions were performed in a final volume of 50 μL in 96-well plates (Phenix Research Products, Hayward, CA) using ABI 9700 (Applied Biosystems) and PTC-200 (MJ Research) thermal cyclers. Incubation conditions for the amplification reactions were 10 min at 24°C to allow the UDG to digest any carryover contamination from previous reactions, 10 min at 95°C to simultaneously inactivate the UDG and activate the Platinum Taq polymerase, 40 cycles of 45 sec at 94°C, 45 sec at 53°C, and 90 sec at 72°C followed by a single cycle of 5 min at 72°C and finally cooling to 4°C.

Amplicons were visualized by agarose gel electrophoresis and ethidium bromide staining using standard conditions (Sambrook and Russell, 2001). For Iowa infection of HCT-8 and Caco-2 cells, amplicon identity was confirmed following gel electrophoresis by hybridization with a digoxigenin (DIG)-labeled oligonucleotide probe specific to the *C. parvum* hsp70 amplicon (cphsp2475, Rochelle et al., 1997). DIG-labeled probe-amplicon hybrids were detected using an anti-DIG antibody conjugated to alkaline phosphatase and a chemiluminescent substrate (CDP Star; Roche).

Table 3.3
PCR primers and probes used in this investigation

Primer/ Probe	Sequence, 5'-3'	Target gene	Size (bp)	T_a*	Reference
cphsp2423F	AAATGGTGAGCAATCCTCTG	hsp70†	361	53°C	Rochelle et al., 1997
cphsp2764R	CTTGCTGCTCTTACCAGTAC				
cphsp2475‡	CCATTATCACTCGGTTTAGA				
F	TCTGGCTATTGGGAACTTG	β-tubulin	618	60°C	Rochelle et al., 1999b
R	CACCTCCCAAAGAGTGAGTG				
F	GTAGATAATGGAAGAGATTGTG	COWP	553	56°C	Spano et al., 1997
R	GGACTGAAATACAGGCATTATCTTG				
cry1§	CGGTTATCCATGTAAGTAAAG	18S rRNA	NA	58°C	Vesey et al., 1998

* Primer annealing temperature.
† Heat shock protein 70 kDa gene.
‡ Oligonucleotide probe used for amplicon confirmation.
§ Probe used for in-situ hybridization.

Isolate	Host	Forward primer (5' – 3')	Reverse primer (5' – 3')
		AAATGGTGAGCAATCCTCTG	CTTGCTGCTCTTACCAGTAC
C. parvum Iowa	bovine	....	....
C. parvum Moredun	cervine	....	....
C. parvum KSU-1	bovine	....	....
C. parvum	Hungarian bovine	....	....
C. parvum	human	....	....
C. parvum	monkey	....	....
C. parvum	ferret	....	....
C. parvum	pig	...C.....A..G..T....	......A.............
C. hominis	human	....	..................G.
C. meleagridis	human	....	....
C. wrairi	guinea pig	....	....
Cryptosporidium sp.	opossum	...............T....	....
Cryptosporidium sp.	red kangaroo	...............T....	....
C. baileyi	quail	......A..A.....A..A.	T..A................
C. muris	mouse	.........A..G..A..A.	...............G....

Figure 3.5 Alignment of hsp70 PCR primers used in this investigation with hsp70 sequences from different isolates and species of *Cryptosporidium*. Identical bases are indicated by dots.

IFA Detection of Infection in Cell Cultures

Infections in MDCK cells were detected using an immunofluorescence assay (IFA). Slide chambers were removed at 48 h post-infection and monolayers were washed with PBS, fixed in Bouin's solution and decolorized with 70% ethanol. Monolayers were washed with PBS and stained with two *C. parvum*-specific monoclonal antibodies. The first monoclonal antibody, 3C3C, labeled all intracellular life cycle stages and was conjugated to Cy3 (Arrowood, 1988). The second antibody, OW50, was conjugated to FITC and labeled oocysts, allowing for detection of adherent oocysts. Antibody-labeled culture chambers were examined using epifluorescence microscopy at 400× magnification using an Olympus BX50 microscope (Olympus America Inc., Lake Success, NY). The entire chamber area was examined for the presence or absence of intracellular parasite stages. As with the RT-PCR detection methodology, IFA was used to simply score the presence or absence of infection in each well of cell culture.

Quality Control of Cell Culture and RT-PCR Procedures

At MWD, all handling, maintenance and culturing of stock and uninfected cells was performed in dedicated cell culture facilities equipped with biological safety cabinets, laminar flow hoods, and cell culture dedicated incubators. Uninfected and *C. parvum*-infected cell monolayers were maintained in separate incubators in different rooms to minimize the risk of false positive infections. Incubators were cleaned weekly by wiping with 70% ethanol and removable shelves were autoclaved. The water in the humidifying tray was replaced weekly with sterile water containing a fungicide. The temperature and CO_2 levels of all incubators were monitored and recorded daily and corrective action taken when necessary. Stock cell lines were stored frozen in liquid nitrogen; liquid nitrogen levels were monitored weekly. Pipets were calibrated annually by Rainin Instruments (Oakland, CA).

Routine quality control (QC) procedures for the cell culturing facility included assessing the sterility of each 100 mL bottle of prepared media and supplements by inoculating Sabaroud-Dextrose agar, blood agar, brain heart infusion broth, and thioglycollate broth (Hardy Diagnostics, Santa Maria, CA). Approximately 5% of all the prepared cell culture media was used for QC purposes. If any batch of prepared cell culture media tested positive for bacterial or fungal growth, it was discarded along with any cells that had been grown in it. At each cell split (twice per week), the cells were observed microscopically to monitor their appearance and performance.

Cell culture log sheets recorded cell line information, passage number, date frozen, date thawed, type of media, size/type of flask, and lot numbers for FBS, trypsin, antibiotics, and all other media components. The logs provided a fully cross-referenced database that allowed any media component or chemical used for any cell culture experiment to be traced to the original manufacturer, lot number, and date received.

Each experiment to measure oocyst infectivity or inactivation included at least two uninfected cell culture wells as negative controls, monolayers inoculated with 1,000 heat killed oocysts (70°C for 30 min) as infection and carryover controls, and monolayers inoculated with approximately 1,000 untreated oocysts as positive controls. Quality control also included tracking all reagents used for the RT-PCR detection assays through assigned lot numbers. Negative amplification controls included omitting the reverse transcriptase from the RT reaction to ensure that amplified product originated from RNA and not DNA, and omitting template from both RT and PCR reactions to ensure amplified product was not the result of laboratory contamination. For selected experiments, extracted RNA was also amplified by RT-PCR using primers specific for a 600 bp fragment of the human glyceraldehyde phosphate dehydrogenase (GAPDH) gene (Stratagene, La Jolla, CA). This is a "house-keeping" gene which is expressed by all human cells (HCT-8 are human cells) and its amplification serves as an internal control of the RNA extraction and RT-PCR procedures. The GAPDH amplicon should be present in all samples at the same signal strength regardless of the oocyst challenge dose applied to the cells or the level of inactivation of oocysts.

Detection of Infection by In-Situ Hybridization

Infection in HCT-8 cell monolayers was also detected using a colorimetric in-situ hybridization (CISH) procedure (Rochelle et al., 2001b). In these instances, cell monolayers were grown in single-well chamber slides. Briefly, cell culture medium was removed from infected cells and monolayers were incubated for 5 min at room temperature in 2 mL of DEPC-treated PBS containing 0.2 μl/mL RNase inhibitor (PE Applied Biosystems, Foster City, CA) followed by DEPC-PBS for a further 5 min. Cells were fixed by incubation in 2 mL of 4% paraformaldehyde (prepared in DEPC-PBS) for 1 h at room temperature. Following fixation, cells were rinsed three times in DEPC-PBS, overlaid with 1 mL of DEPC-PBS, incubated at 60°C for 15 min, and rinsed twice with DEPC-PBS. Cellular membranes were then permeabilized by incubation in 2 mL of proteinase K (10 μg/mL in 50 mM Tris-HCl, pH 7.6) for 12 min at 37 °C, rinsed in DEPC-PBS, refixed in 2 mL of 2% paraformaldehyde at room temperature for 10 min, rinsed in DEPC-PBS, and finally dehydrated in 75% ethanol for 5 min followed by 100% ethanol for 5 min. The plastic enclosure and gasket of the 1-well chamber slides were removed and the slides were allowed to dry. The cells were overlaid with 175 μL of hybridization solution (0.9 M NaCl, 20 mM Tris-HCl, pH 7.5, 0.1% SDS, 4 U/mL RNase inhibitor) containing 70 nM of the cry1 oligonucleotide probe (Table 3.3) and heated to 99°C for 15 sec and then incubated at 45°C for 3 h in a humidified chamber. Slides were then incubated in pre-warmed 5 × SSPE (1 × SSPE is 150 mM NaCl, 10 mM monobasic sodium phosphate, 1 mM EDTA, pH 7.4) at 37°C for 10 min, incubated in 2.5 × SSPE for 10 min at 37°C, and then in 2 mL of 0.1 M Tris-HCl, pH 7.4, 0.15 M NaCl containing 1% blocking reagent (InnoGennex, San Ramon, CA) at room temperature for 10 min. Cells were incubated in 100 μL of anti-digoxigenin alkaline phosphatase conjugate (Roche Molecular Biochemicals; 1.5 U in 100 mM Tris-HCl, pH 7.5, 150 mM NaCl, 3% BSA, 0.1% Triton X-100) at 28°C for 45 min on a rocking platform, rinsed twice with Tris buffered saline, and incubated in substrate buffer (100 mM Tris-HCl, pH 9.5, 50 mM $MgCl_2$, 100 mM NaCl) for 5 min. Cells were then overlaid with 175 μL of nitroblue tetrazolium/5-bromo-4-chloro-3-indoyl phosphate substrate solution (Vector Laboratories, Burlingame, CA) containing levamisole prepared according to the manufacturer's instructions (Vector Laboratories) and incubated under a 22 × 50 mm glass coverslip for 12 – 18 h in a humidified, dark chamber at 28°C. Finally, slides were rinsed in deionized water, dried, mounted, and examined by light microscopy. Infection was quantified by enumeration of the number of infectious foci that developed in the monolayer and the average number of parasite developmental stages per foci.

OOCYST PRETREATMENT PRIOR TO CELL CULTURE INFECTIONS

For some experiments, oocysts were incubated in 0.75% sodium taurocholate (prepared in RPMI medium) at 37°C for 10 – 30 min to trigger excystation and washed with PBS prior to inoculation of monolayers.

AGING OOCYSTS IN ENVIRONMENTAL WATER SAMPLES

Oocysts of the Iowa isolate were seeded into 10 L of untreated plant influent from one of Metropolitan's filtration plants at a concentration of 2×10^4/mL. The in-vitro excystation efficiency for this batch of oocysts was 95.8%. The characteristics of this water sample at the time of collection were: temperature, 10.8°C; turbidity, 0.67 NTU; total dissolved solids, 314 mg/L; total suspended solids, 0.6 mg/L; pH 8.4; conductivity, 569 μS/cm; total hardness, 118 mg/L; dissolved oxygen, 10.9 mg/L; and free chlorine residual, 0 mg/L. The seeded water was stored in the basement of Metropolitan's demonstration-scale treatment plant so that it was exposed to ambient fluctuations in temperature; the temperature range of the oocyst suspension during the 35-day period of the experiment was 12 – 19°C, as measured with a minimum-maximum recording digital thermometer. A similar carboy of water containing oocysts was also stored at 4°C during the duration of the test to serve as a control. The ID_{50} of the fresh oocysts was determined by HCT-8 cell culture at the start of the test and for the control oocysts at the end of the test. The ID_{50} of oocysts maintained at environmental conditions was determined following storage for 35 days. Oocysts were recovered from the untreated source water by using USEPA Method 1623 (USEPA, 1999), disassociated from paramagnetic beads in 0.1 M HCl and neutralized with NaOH in solution rather than on a slide, and diluted to the required challenge doses. Prior to recovery by filtration, an aliquot of the oocyst suspension was removed and oocysts recovered by centrifugation. The infectivity of the recovered oocysts was then assessed by HCT-8 cell culture and the CD-1 mouse assay. The oocysts that were recovered directly by centrifugation were incubated in 10% bleach (0.5% sodium hypochlorite in sterile deionized water) for 10 min at 4°C and then washed by repeated centrifugation at $4{,}000 \times g$ and resuspension in sterile PBS prior to inoculation of cells or mice. The intent of bleach treatment was to surface sterilize the oocysts to eliminate bacterial contaminants that may have been present in the source water.

QUANTIFICATION OF INFECTIVITY

To ensure that the results obtained with the different infectivity models were directly comparable, the same procedure was used to quantify the level of infection in all models. For mouse assays, each challenge dose of oocysts was inoculated into 3 – 6 replicate cages of mice, each cage containing 9 – 12 mice per cage. Similarly, for cell culture assays, each challenge dose was inoculated into 2 – 7 replicate sets of cell monolayers, each set containing 6 – 10 monolayers per set. All infectivity results were expressed as the percent of cell culture monolayers or animals that became infected at each oocyst dose. Dose response curves were constructed by fitting a least squares regression to a plot of oocyst challenge dose against a logistic transformation of the proportional infectivity (Korich et al., 2000a). The response logit (*RL*) was defined as:

$$RL = Log_e \frac{(P/100)}{[1-(P/100)]} \quad (3)$$

where Log_e = the natural logarithm
P = the percentage of infected wells or mice

The ID_{50} values were calculated by solving the logistic regression for a logit value of zero. The logit function does not allow logarithmic transformations of 0 and 100% responses so any set of monolayers or cage of animals that generated 0 or 100% infection (P = 0% or 100%) was excluded from the logit response model. The logit dose response method ensures linearity of dose response curves and is an accepted analytical tool for expressing parasite infectivity; it has been used previously for inter-laboratory comparisons of *C. parvum* infectivity in mice (Korich et al., 2000a).

Statistical Analyses

Data generated in Phase 1 of the project comparing mouse and cell culture infectivity models for measuring infectivity were analyzed for normality using the Shapiro-Wilk test for sample sizes of less than 2,000. Out of 70 datasets analyzed (five isolates in mouse and cell culture infectivity models), the data were normally distributed ($P>0.05$) in 87% of cases. Since the majority of the data were normally distributed, a parametric two-tailed t-test was used to compare infectivity in different models. A two-tailed test was used because there was no a-priori expectation that one method would generate consistently higher levels of infection than any other method. A parametric Chi-square test was used to determine whether there were significant differences in the percent infectivity resulting from different doses of each isolate. Dose response relationships were compared using one-way analysis of covariance (ANCOVA) on percent infectivity values to determine whether there were significant differences between the different infectivity models for each isolate. In addition, the means and standard deviations of replicate percent infectivity values for each infection detection method, *C. parvum* isolate, and oocyst dose were used as inputs for a standard power analysis (α = 0.05, β = 0.80) to determine the number of replicates of mice or cell culture wells needed to allow detection of specific differences in percent infectivity.

UV DISINFECTION OF OOCYSTS

Oocysts were exposed to UV light using a medium pressure UV lamp mounted above a 6.4 cm inside diameter collimating tube (Calgon Carbon Corporation, Pittsburgh, PA). Oocysts were suspended at a concentration of $1 \times 10^5 - 5 \times 10^6$/mL in 10 mL of PBS or treatment plant effluent in a 60-mm polymethylpentene petri dish and exposed to a variety of UV dosages to achieve a range of inactivation levels. The samples were continuously mixed using a stir bar and magnetic stirrer during UV exposure. Oocysts were exposed for variable lengths of time, at a constant distance, from a constant intensity UV light to achieve the different UV dosages (mJ/cm^2).

Incident UV irradiance was measured using an IL1700 radiometer (International Light, Newburyport, MA) with mathematical correction and confirmed utilizing chemical actinometry to calibrate the dosages delivered to the oocyst suspensions. The incident UV irradiance was measured at the center of the liquid surface of the sample. By making appropriate measurements at positions away from the center, the average incident UV irradiance across the top of the petri dish was obtained and the average UV dose in the 10 mL sample was calculated using the integration of the Beer-Lambert Law through the depth of the liquid sample, adapted for polychromatic radiation. A more detailed description of the UV apparatus and the methods used to measure UV dosages has been published previously (Mofidi et al., 2001).

Quantifying Inactivation

Each challenge dose of untreated control oocysts or UV-exposed oocysts was inoculated into at least six wells of HCT-8 monolayers or 10 CD-1 mice and the infectivity data was transformed using the logit response described above. To determine the level of inactivation, dose response curves were constructed for untreated control oocysts and oocysts exposed to UV light and the level of inactivation was then determined by comparing the ID_{50} of exposed oocysts to the control, untreated oocysts. Inactivation was expressed as the logarithm (base 10) of the increase in ID_{50} following exposure to UV:

$$Inactivation = Log_{10}\left(ID_{50}^{C}\right) - Log_{10}\left(ID_{50}^{UV}\right) \qquad (4)$$

where ID_{50}^{C} = the 50% infectious dose for control oocysts

ID_{50}^{UV} = the 50% infectious dose for UV exposed oocysts

OZONE DISINFECTION OF OOCYSTS

Ozone disinfection experiments were performed in three types of reactor: simple batch, semi-batch, and a single continuously stirred tank reactor (CSTR). The majority of experiments were performed using HCT-8 cell culture combined with RT-PCR to measure infectivity and inactivation. Some experiments were also performed using the CD-1 mouse assay in parallel with cell culture, and the MDCK cell culture/IFA assay was used for a limited number of experiments.

With one exception early in the study, all ozone inactivation experiments were performed with the Iowa isolate of *C. parvum*. Oocysts were propagated and purified using the methods described in Chapter 3 and were less than 30 days old (post-shedding) at the time they were used for disinfection experiments. Upon receipt at MWD, oocysts were recovered from the saline antibiotic solution by centrifugation and resuspended in 0.05 M phosphate buffer (pH 7). They were enumerated using a hemacytometer and introduced into the ozone reactors at the required concentrations (typically $1 \times 10^5 - 1 \times 10^6$ oocysts/mL). Ozone disinfection experiments were performed in either 0.05 M phosphate buffer (pH 7) or untreated source water from Metropolitan's treatment plants, at ambient pH (Table 3.4). Ozone was produced from 100% oxygen by an Osmonics V5-0 generator (Minnetonka, MN).

Following the required exposure periods, residual ozone in the oocyst samples was quenched by the addition of sodium thiosulfate to a final concentration of 0.02%. Oocysts were then diluted in PBS, enumerated by hemacytometer and IFA, and inoculated onto cell monolayers or into mice to measure infectivity. Inactivation was assessed by comparing the ID_{50} values for control and ozone exposed oocysts and by solving the logistic dose response for individual infectivity values for ozone exposed oocysts. Although oocysts were washed and resuspended in PBS following ozone exposure, the effect of adding sodium thiosulfate to cell monolayers was tested. Non-ozonated oocysts treated with 2.5% sodium thiosulfate were centrifuged and resuspended in PBS and then inoculated onto HCT-8 cells. There was no difference in the level of infectivity of these isolates compared to control oocysts indicating that any residual sodium thiosulfate that may carry through with the oocysts will not have a detrimental effect of infectivity in cell culture.

Influence of Ozone on Immunofluorescent Staining of Oocysts

Since ozonated oocysts were enumerated by immunofluorescence microscopy prior to inoculating cell cultures, the effect of ozone on immunofluorescent staining of oocysts was evaluated. Oocysts of the Iowa isolate were suspended in 0.05 M phosphate buffer (pH 7.4) and exposed to ozone in a 20 mL volume batch reactor; ozone was applied to achieve an initial concentration of 4.8 mg/L. The oocyst concentration was 2×10^3/mL and they were 16 days old at the time of testing. Following the exposure periods required to achieve CT values of 4.8, 9.6, and 28.8 mg.min/L, residual ozone was neutralized by the addition of sodium thiosulfate and five replicate 100 μL aliquots were removed from the reactor, applied to 2-well microscope slides and stained with FITC-labeled anti-*Cryptosporidium* antibody (Waterborne, Inc.). Oocysts were enumerated by epifluorescence microscopy. The average number of oocysts in 100 μL aliquots of untreated control oocysts was 59 ± 8.5 (mean ± CV) and 58.7 ± 19.8 for ozone exposed oocysts. One-way analysis of variance confirmed that there was no significant difference between the staining efficiency of ozone-exposed oocysts and untreated oocysts (P = 0.88, 95% confidence level, N = 5 for each treatment).

Table 3.4
Raw water quality

		Mean value	
Parameter	Units	SPW*	CRW† blend
Temperature	°C	16	17
pH	--	8.0	8.3
Total organic carbon (TOC)	mg/L	2.9	2.8
Specific ultraviolet absorbance‡	L/mg•m	3.0	1.5
Alkalinity	mg/L as $CaCO_3$	70	130
Turbidity	NTU	2.3	1.1

* California State Project Water. † Colorado River Water.
‡ SUVA = 100 × UV/TOC.

Continuously Stirred Tank Reactor (CSTR)

The CSTR consisted of a 500 mL gas-washing bottle modified to provide continuous water and gas flow through the reactor. Temperature was controlled using an in-line heat exchanger and a water jacket surrounding the contactor. Ozone-enriched oxygen (approximately 3% ozone) was introduced through a porous glass frit at the bottom of the reactor and collected in the reactor headspace. Water was pumped from a 20 L feed reservoir through the reactor using a peristaltic pump. Two, 20 L raw water feed reservoirs were used—one containing a suspension of *C. parvum* oocysts and the second serving as a control. Both reservoirs were connected to a reactor feed pump and a multi-port valve facilitated switching between water sources. At the beginning of each experiment, target disinfection levels were established using the control water source. Once an appropriate disinfection level was established the *C. parvum*-spiked water was then introduced into the reactor. A schematic illustration of the ozone contactor is shown in Figure 3.6. The flow rate of the process water was varied to achieve a range of CT values while keeping the applied ozone dose within practical limits. Reactor detention times were in the range 15 – 30 min. The oocyst concentration in the feed reservoir was typically 5×10^4/mL in a 20 L volume and sample volumes ranging from 45 to 150 mL were collected through a sample port for each inactivation level (the volume collected depended on whether infectivity was being done at just one laboratory or all three of the analysis laboratories involved in this study).

Dissolved ozone was measured using a modification of the indigo method as described in Standard Method 4500-O_3 (Clesceri et al., 1998). Samples were collected using an indigo-filled syringe and quantified using a UV/VIS spectrophotometer. To assess the effects of process variation, three sample replicates were collected at different times. Both applied and transferred ozone doses were measured during the experiments. CT was calculated as the product of hydraulic detention time, dissolved ozone residual and hydraulic performance using two methods:

$$CT_{10} = C \times HRT_{10} \qquad (5)$$

and

$$CT = C \times HRT \qquad (6)$$

Where C = Dissolved ozone concentration in reactor (mg/L)

HRT_{10} = Hydraulic retention time (min) × t_{10}/t_θ ratio (min; = 0.1 for ideal CSTR)

HRT = Hydraulic retention time (min) = Reactor volume (mL)/Flow rate (mL/min)

For an ideal single-CSTR, as used in this study, the t_{10}/t ratio equals 0.1. The actual t_{10}/t ratio for the reactor used in this study was verified through tracer tests (see Figure 5.6).

CSTR experiments were performed at 5°C and 15°C, in 0.05 M phosphate buffer (pH 7) and untreated source water, using just the Iowa isolate. Eleven experiments were performed generating 151 cell culture records and 96 mouse assay records as summarized in Table 3.5.

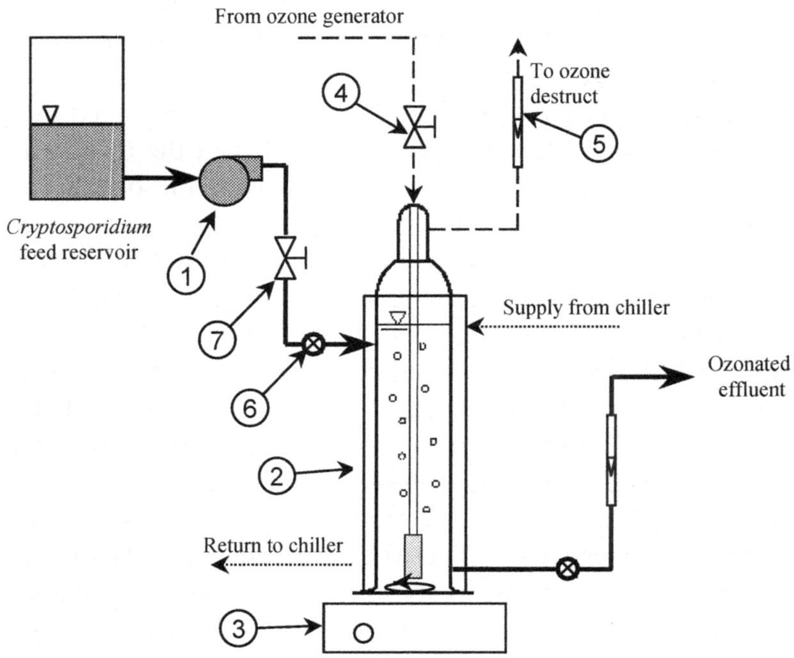

Figure 3.6 Schematic illustration of bench-scale continuously stirred tank ozone contactor. Liquid and gas streams are indicated by: – – – –, Ozone stream; ——, Process stream; ·······, Cooling water stream. The various components were: 1) Peristaltic feed pump (500 mL/min capacity); 2) 500 mL jacketed gas wash bottle with inlet and outlet ports; 3) Magnetic stirrer; 4) Ozone gas flow regulation valve; 5) Rotameter for liquid or gas flow; 6) Liquid sample tap; 7) Water flow regulation valve.

Table 3.5
Summary of continuously stirred tank reactor ozone experiments

Method	Matrix	Temp. (°C)	Lab.	CT (mg.min/L)	Oocyst dose*
Cells	SPW	15	MWD	0	74,762
				1.9	88,908,10000
				4.8	64,675,10000
				5.4	78,818,10000
				7.1	70,699,10000
	0.05M PO_4	15	MWD	0	86,784
				4.5	75,748,9000
				0	45,93,489
				4.1	84,456,897
				11.1	81,793,7700
				31.2	71,814,8500,
				0	21,41,77,159,174,447
				3.9	38,201,1113
				10.6	31,163,825
				35.7	23,123,578
	0.05M PO_4	5	MWD	0	45,235,1131
				6	41,221,1148
				15.4	243,1240
				26.8	1188,6176,
	0.05M PO_4	5	EU	0	50,250
				6	50,250,1250
				15.4	250,1250
	0.05M PO_4	15	MWD	0	18,36,73
				17.2	40,402,804
				70.5	107,536,5360
Mice	SPW	15	UA	0	25,128
				4.8	117, 1050,2800
				5.4	118,1300
				7.1	106,1150,6500
	0.05M PO_4	15	UA	0	50,100,500
				4.1	100,500,10000
				11.1	100,1000,10000
				31.2	100,1000,10000
				0	40,200,1000
				3.9	40,200,1000
				10.6	40,200,1000
				35.7	40,200,1000

(continued)

Table 3.5 (Continued)

Method	Matrix	Temp. (°C)	Lab.	CT (mg.min/L)	Oocyst dose*
Mice	0.05M PO_4	5	UA	0	50,100,150
				6	50,250,1250,
				15.4	250,1250,6250

* Mean oocyst dose inoculated into each mouse or cell culture. Multiple oocyst doses from each CT condition were inoculated so that dose response curves could be constructed for untreated and ozone treated oocysts.

Abbreviations: EU, Emory University.
MWD, Metropolitan Water District.
PO_4, phosphate buffer.
SPW, State Project Water (untreated source water).
UA, University of Arizona.

Semi-Batch Reactor

The semi-batch reactor was constructed out of borosilicate glass with an effective volume of 2 L. The ozone contact chamber was surrounded by an integral glass jacket to allow circulation of ethylene glycol through an external chiller for temperature control (Figure 3.7). Samples were collected from a port at the bottom of the vessel or through ports in the reactor lid. A stainless steel clamp secured the reactor lid to the body with a Teflon O-ring providing the seal. Ozone enriched oxygen was delivered by sparging through an array of four medium porosity glass frits. Prior to each test the reactor was sanitized by ozonating at a relatively high dose (up to 10 mg/L) for extended periods (1 – 3 h). The reactor was then drained and flushed with 2 L of ultrapure water. The reactor contents and rinse water were autoclaved before disposal. While this was not the most efficient sanitization procedure it was more practical than autoclaving. Initially autoclave sterilization was used but the complicated structure of the glass vessel led to differential heating and cooling during autoclaving resulting in damage to the reactor.

Once sterilized, 2 L of sample water was added to the reactor. Ozone gas was added and grab samples (~5 mL volumes) were collected and analyzed for dissolved ozone using the indigo method. Once the dissolved ozone concentration had reached a steady state (usually after 10 – 20 min), oocysts were added using a 10-mL gas tight syringe connected to a Luer type adapter in the reactor lid. The syringe was then flushed once with sterile 0.05 M phosphate buffer, which was also added to the reactor. Dissolved ozone samples were collected using a timer-controlled peristaltic pump. Prior to sample collection, the 1/8" sample line was purged for 30 sec and the sample was then collected into a 15 mL polypropylene vial containing 1 – 3 mL of indigo trisulfonate. Samples for measuring infectivity were collected using a sterile 50 mL gas-tight syringe. Collecting samples rapidly (within ~30 sec) is crucial when using the semi-batch method due to the time dependence of CT. Once collected in the syringe, samples were immediately transferred to a 50 mL polypropylene vial containing 0.5 mL of sterile 0.02 N sodium thiosulfate to quench the dissolved ozone residual. CT was determined by continually measuring dissolved ozone concentration (at 1 – 3 min intervals) over the course of the experiment. Time measurement began when oocysts were inoculated. CT was calculated according to the following expression:

$$CT = \sum_{i=2}^{n} \frac{C_i + C_{i-1}}{2} \times (t_i - t_{i-1}) \qquad (7)$$

Where CT	=	Ozone exposure (mg/L x min)
n	=	Number of dissolved ozone samples collected over the entire test
C	=	Dissolved ozone concentration (mg/L)
t	=	Time (min).

Semi-batch reactor experiments were performed at 1°C, 5°C, 10°C and 15°C, in 0.05 M phosphate buffer (pH 7), two untreated source waters and a blend of the two waters, using just the Iowa isolate. A total of 501 data records were generated for the semi-batch experiments as summarized in Tables 3.6 and 3.7.

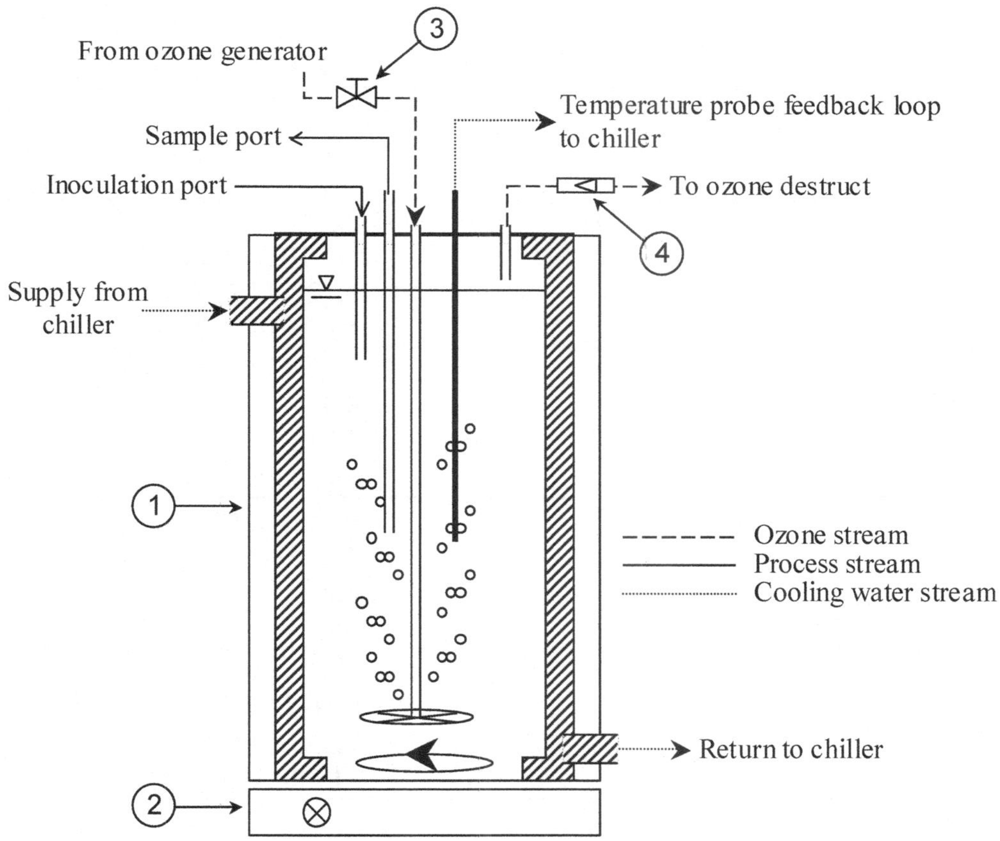

Figure 3.7. Schematic of semi-batch ozone reactor. Liquid and gas streams are indicated by: – – – –, Ozone stream; ——, Process stream; ········, Cooling water stream (the coolant jacket is indicated by diagonal hatching). The various components were: 1) 2 L glass reactor vessel with inlet and outlet ports; 2) Magnetic stirrer; 3) Ozone gas flow regulation valve; 4) Rotameter for liquid or gas flow.

Table 3.6
Summary of semi-batch experiments involving cells

Matrix	Temp. (°C)	Lab	CT (mg.min/L)	Oocyst dose*
0.05M PO_4	15	MWD	0	38,76,198
			3.6	73,397
			11	385,712
			19.6	760,4070
			0	48,97,261
			7	19,36,217
			20.4	76,204,789
0.05M PO_4	15	EU	0	50,100
			7	100,250,500
			20.4	100,250
0.05M PO_4	5	MWD	0	40,90,220
			7.9	49,104,477
			16.2	386,812
			30.4	828,4068,10000,23000
0.05M PO_4	15	MWD	0	19,67,133
			2.9	19,61
			4.5	18,62,129
CRW pH = 8.4	5	MWD	0	22,24,41,44,87,232,235
			7.3	3,7,10,25,86,221
			21.8	12,27,42,126,189
			36.1	15,31,143,228,445
CRW pH = 8.24	5	MWD	0	9,23,43,86
			16.2	8,41
			30.4	9,41,254
			50.3	58,274,555
CRW pH = 8.18	5	MWD	0	6,19,50,105
			28.3	11,25,60
			43.3	8,89
			79.8	852
CRW/SPW blend (50/50)	5	MWD	0	12,16,21
			3.6	9,16,25,70,299
			8.7	19,53,209
			17.5	71,280,707,1544

(continued)

Table 3.6 (Continued)

Matrix	Temp. (°C)	Lab	CT	Oocyst dose*
0.05M PO_4	1	MWD	0	11,28,57
			36.6	103,515
			56.1	521,1041
			75.1	962,4812
			0	12,29,59,118
			15.9	12,59,588
			31.6	107,534,1068,5342
			47.9	131,1306,13057
			62.8	6541
			0	7,43,86
			5.5	6,32,63,316
			9.5	7,35,69,347
			20.2	33,67,333,667
			0	11,48,53,96,106
			9.9	40,81
			18.8	52,104,522
0.05M PO_4	1	EU	0	50,100
			18.8	25,50,100
CRW pH = 8.3	5	MWD	0	43,73,353
			22.7	35,70,380
			45.3	413,825,5000
0.05M PO_4	10	MWD	0	50,100,250
			11.4	25,50,250
			22.2	100,250,500

* Mean oocyst dose inoculated onto each cell monolayer. Multiple oocyst doses from each CT condition were inoculated so that dose response curves could be constructed for untreated and ozone treated oocysts.

Abbreviations: CRW, Colorado River Water (untreated source water).
EU, Emory University.
MWD, Metropolitan Water District.
PO_4, phosphate buffer.
SPW, State Project Water (untreated source water).

Batch Reactor Experiments

Batch reactors were 250 mL Pyrex glass bottles containing a suspension of oocysts to which an aliquot of an ozone stock solution was added (stock concentration = 9 mg/L). Samples were removed and quenched with sodium thiosulfate at appropriate intervals. The stock solution was generated by bubbling ozone through 0.05 M phosphate buffer (pH 7). Batch experiments were performed in 0.05M phosphate buffer (pH 7), at 15°C and 20°C, with the Moredun and Iowa isolates of *C. parvum*. A total of 7 experiments generated 83 cell culture records and 37 mice records, as summarized in Table 3.8.

Table 3.7
Summary of semi-batch experiments involving mice

Matrix	Temp. (°C)	Lab.	CT (mg.min/L)	Oocyst dose*
0.05M PO_4	15	UA	0	51,119
			7	47,97,488,1100
			20.4	136,575,5875
0.05M PO_4	5	UA	0	50,104,
			16.2	638,4833,10708
			30.4	9583,26958
CRW pH = 8.4	5	UA	0	48,110
			7.3	64,258
			21.8	533,1275
			36.1	5150
CRW pH = 8.18	5	UA	0	35,53
			28.3	11,49
			43.3	13,57
CRW/SPW blend (50/50)	5	UA	0	109
			3.6	113
			8.7	93,279
			17.5	227,1075
0.05M PO_4	1	UA	0	30,165
			47.9	1100,10375
			62.8	1300,10625
			0	52,108
			18.8	50,111,500
CRW pH = 8.3	5	UA	0	150
			22.7	700,1350
			45.3	1000,4875

* Mean oocyst dose inoculated into each mouse.

Abbreviations: CRW, Colorado River Water (untreated source water).
PO_4, phosphate buffer.
SPW, State Project Water (untreated source water).
UA, University of Arizona.

Table 3.8
Summary of batch ozone experiments

Method	Isolate	Temp. (°C)	Lab.	CT (mg.min/L)	Oocyst dose*
Cells	Moredun	20	MWD	0	70,710
				1.34	108,930,9800
				2.5	940,9200
			EU	0	100,1000
				1.34	100,1000,10000
				2.5	100,1000,10000
	Iowa	15	MWD	0	45,92,207,409
				16	74,351,2507
				220	7200,29400
			EU	0	50,100,400
				16	100,2000
				220	400,8000
			MWD	0	33,152,960
				9.8	25,155,236,1085,1160,4537,6950
Mice	Moredun	20	UA	0	100,1000,
				1.34	100,1000,10000
				2.5	1000,10000
	Iowa	15	UA	0	50,100,200,400
				16	100,400,2000
				220	8000,32000

* Mean oocyst dose inoculated into each mouse or cell culture.
Abbreviations: EU, Emory University.
MWD, Metropolitan Water District.
UA, University of Arizona.

OZONE INACTIVATION OF *BACILLUS SUBTILIS* SPORES

Cultures of *Bacillus subtilis* (ATCC 6633) were grown in 500 mL flasks containing 50 mL of half-strength nutrient agar for 96 h at 35°C. Cells were harvested by scraping the resulting culture off the agar and resuspended in PBS. Cells were washed three times by centrifugation and resuspension in PBS. Vegetative cells were killed by heating the suspension at 80°C for 20 min, followed by cooling to 4°C. Spores were harvested by centrifugation at 4,400 × *g* for 15 min, washed by resuspension in PBS and centrifugation, and finally resuspended in 10 mL of PBS.

Spores were introduced into the 2 L semi-batch reactor containing untreated, filter sterilized source water and exposed to ozone at 1°C, 5°C, and 15°C using the same procedures and equipment as used for *C. parvum*. Source water was passed through 0.2 μm porosity membrane filters for these *B. subtilis* experiments to remove interfering indigenous microbes.

The sample matrix for experiments performed at 1°C and 15°C was a blend of 80% Colorado River Water (CRW) and 20% California State Project Water (pH 8.4, 0.08 NTU). For the 5°C experiments, the sample matrix was 100% CRW (pH 8.2, 0.05 NTU). Typical characteristics for these natural waters are provided in Table 3.4. Following the required contact time, samples were withdrawn from the reactor and residual ozone was neutralized by the addition of sodium thiosulfate to a final concentration of 20 mM. Exposed and non-exposed spores were enumerated by triplicate serial dilution in PBS, filtering 10 mL samples through 0.45 μm nitrocellulose membrane filters, and incubation of the filters on nutrient agar plates at 35°C for 18 h. The maximum amount of inactivation that could be detected by this assay was 6.8-log_{10}.

CHAPTER 4
RESULTS AND DISCUSSION

PHASE 1: COMPARISON OF CELL CULTURE AND MICE FOR MEASURING INFECTIVITY OF UNTREATED OOCYSTS OF *CRYPTOSPORIDIUM*

The first phase of this project involved comparing the CD-1 mouse model with three cell lines and two cell culture assays for measuring the infectivity of freshly produced oocysts from five isolates of *C. parvum*. A comparison was also made between CD-1 mice and a single cell line for measuring the infectivity of a *C. hominis* isolate. Infection in CD-1 mice was detected by staining tissue sections with hematoxylin and eosin followed by microscopic observation. Infection in MDCK cell cultures was detected by application of a fluorescently labeled sporozoite-specific antibody and fluorescence microscopy. This technique is similar to that used for routine detection of oocysts in water. Infection in HCT-8 and Caco-2 cell cultures was detected by RT-PCR targeting mRNA transcribed from a *C. parvum*-specific hsp70 gene. Messenger RNA was selected as the target for detection since it is actively produced by the parasite as it develops in cell culture, but is rapidly degraded if oocysts are incapable of initiating infection. Thus, the risk of false-positive detection of dead or non-infectious oocysts is greatly reduced. PCR-based methods which detect DNA may detect dead oocysts which are simply located on the cell monolayer but not actively infecting. DNA is far more stable than mRNA and may persist for extended periods inside dead oocysts.

COMPARISON OF CELL CULTURE AND MICE FOR MEASURING INFECTIVITY OF FRESH OOCYSTS OF *CRYPTOSPORIDIUM PARVUM*

This project was the first large-scale, statistically robust study to compare in-vitro cell culture with a mouse assay for measuring infectivity of multiple isolates of *C. parvum*. The objective of the study was not to maximize the sensitivity of infection in cell culture and then compare this optimized assay with a mouse assay. Instead, the intent was to directly compare the shape of the dose response curves and the level of variability in cell culture assays and mice using oocysts that had been handled and treated in the same way for both types of assay.

Overall, the results in Phase 1 of the project with fresh oocysts from five isolates of *C. parvum* (previously referred to as genotype 2 *C. parvum*) demonstrated good agreement between the CD-1 mouse and cell culture infectivity assays in terms of the shape of dose response curves (Figures 4.1 – 4.4), range of ID_{50} values (Tables 4.1 – 4.4), and the level of variability (Rochelle et al., 2002). The logit dose response model excludes 100% and 0% infectivity results because the natural logarithm of 0/0 and 1/0 cannot be obtained. Therefore, a certain amount of data that was generated by each of the infectivity models was lost from the analysis. For CD-1 mice, out of 120 datapoints generated from 1,551 individual mice, 35 (29%) were either 0% or 100%. Likewise, the proportions of data omitted for the cell culture methods were 15% for HCT-8 cells (126 out of 1,045 individual wells of cell culture), 19% for Caco-2 cells (100 out of 805 wells of cell culture), and 36% for MDCK cells (113 out of 1,005 wells of cells).

Chi-square contingency analyses demonstrated that there were significant differences in the infectivity of different oocyst doses for all isolates and all infectivity models ($P<0.01$) except for the Moredun and Iowa isolates in MDCK cells with infection detected by IFA (P=0.34). This indicated that in these particular instances, increasing oocyst dose did not lead to a significant increase in the level of infectivity. This is illustrated by the data for the Moredun isolate in MDCK cells (Figure 4.4) with a percent infectivity of 75% (logit = 1.0986) at 50 oocysts increasing to just 85% (logit = 1.7346) with 500 oocysts. For all other isolates and infectivity models, an expected dose-response relationship was observed (i.e. increasing oocyst dose leading to an increase in the amount of infection). The lack of a significant dose response in the MDCK/IFA assay for the Moredun and Glasgow isolates was also reflected by the very low ID_{50} values of 1.2 and 1.5 oocysts (Table 4.4) and significant differences in the slopes of the dose response regressions (ANCOVA, P<0.05) compared to the other infectivity models for these two isolates. With these two exceptions, ANCOVA on the data used to generate the dose response curves for the Iowa, Moredun, Maine, and Glasgow isolates demonstrated that the slopes of the regression lines were not significantly different (range of P = 0.11 – 0.98). Two-tailed t-tests on replicate percent infectivity data for each dose of oocysts of each isolate (total of 26 comparisons) demonstrated that HCT-8 cell culture yielded a significantly higher level of infection compared to CD-1 mice in 4 instances (range of P values was <0.01 – 0.04). Conversely, CD-1 mice demonstrated higher levels of infection in 6 instances (P = <0.01 – 0.03). These differences were distributed evenly over all isolates. There was no significant difference between HCT-8 cells and CD-1 mice (P = 0.06 – 0.91) for the remaining 16 comparisons.

The dose response curves in Figures 4.1 – 4.4 were generated by performing the logit transformation on each individual measure of percent infectivity at each oocyst dose without averaging. However, dose response curves can also be generated using the logit transformation on the average percent infectivity from all replicates at each oocyst dose, allowing 0% and 100% infectivity to be included in the calculation of averages, so long as at least one of the replicates did not yield a 0% or 100% result. Both approaches were adopted for calculating the ID_{50} values in Tables 4.1 – 4.4 and in some instances the two methods of calculation led to substantial differences in the ID_{50} (e.g., Moredun in CD-1 mice and Iowa in Caco-2 and MDCK cells).

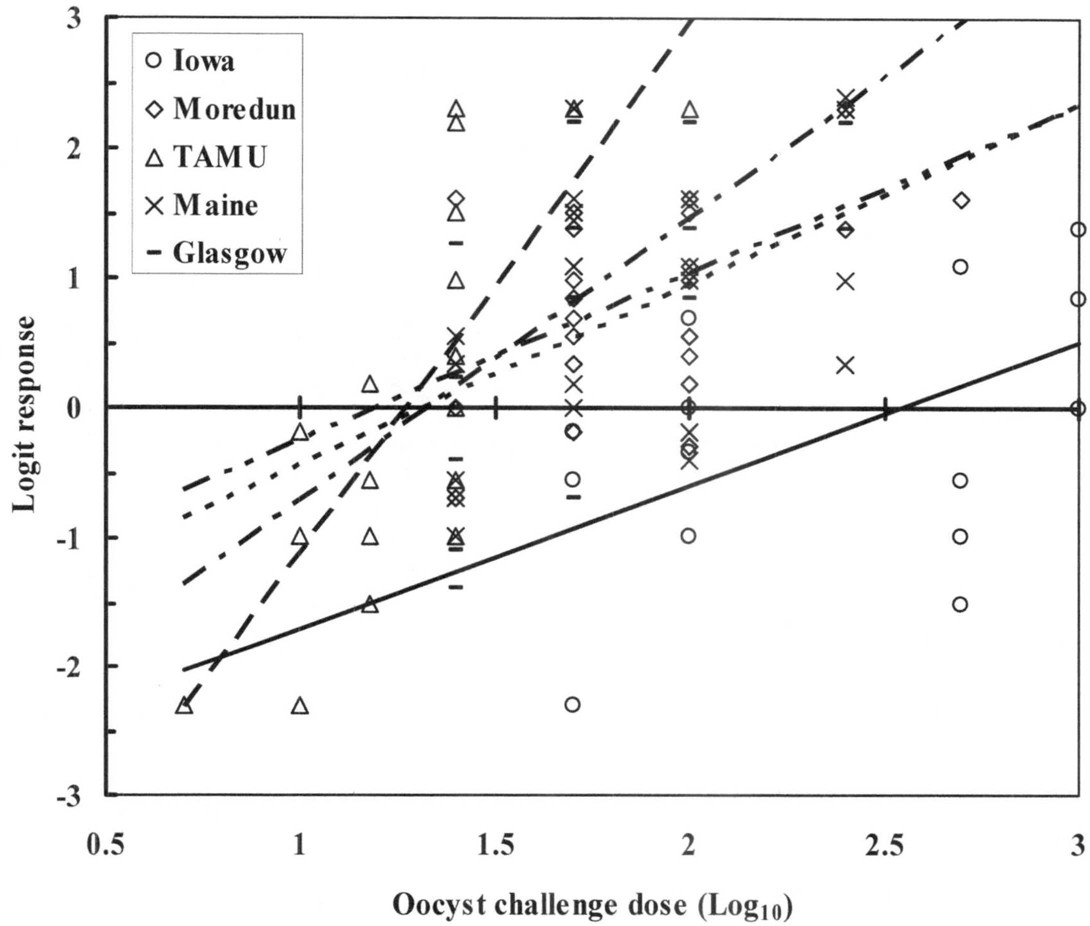

Figure 4.1 Dose response curves for five isolates of *C. parvum* in CD-1 mice. The ID_{50} is the point at which the regression line crosses the oocyst challenge dose axis (logit response = 0). Infectivity was measured for the Iowa (○, —), Moredun (◊, — — — —), TAMU (Δ, — — —), Maine (×, ------), and Glasgow (–,— - —) isolates. Each data point represents the logit transformation of the proportional infectivity for three to six replicate cages of mice with 9 – 12 mice per cage, at each oocyst dose. The graph represents 120 datapoints derived from 1,551 mice.

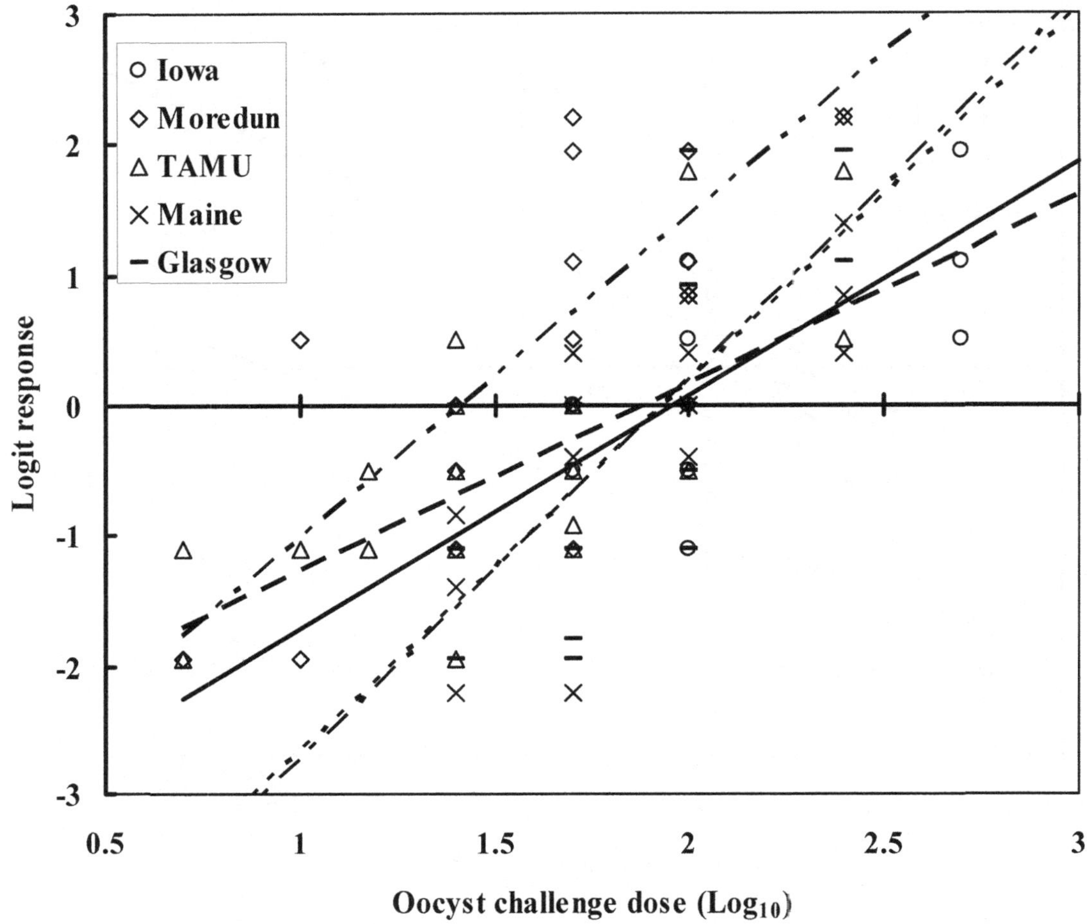

Figure 4.2 Dose response curves for five isolates of *C. parvum* in HCT-8 cells. Infectivity was assessed using an RT-PCR assay. The ID_{50} is the point at which the regression line crosses the oocyst challenge dose axis (logit response = 0). Infectivity was measured for the Iowa (○, ——), Moredun (◊, — — — —), TAMU (Δ, — — —), Maine (×, ------), and Glasgow (–,— — —) isolates. Each data point represents the logit transformation of the proportional infectivity for two to seven replicate sets of cell cultures with 6 – 10 monolayers per replicate. The graph represents 126 datapoints derived from 1,045 individual wells of cell culture.

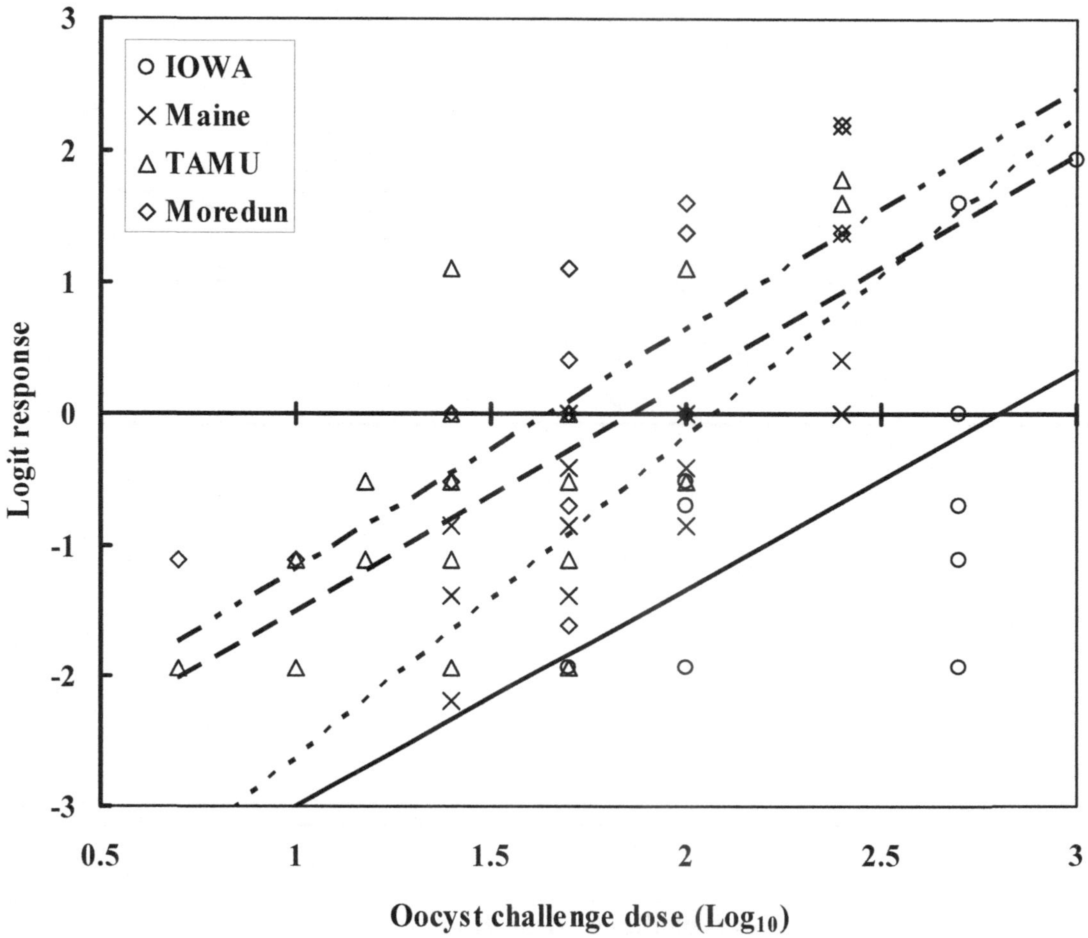

Figure 4.3 Dose response curves for four isolates of *C. parvum* in Caco-2 cells. Infectivity was assessed using an RT-PCR assay. The ID_{50} is the point at which the regression line crosses the oocyst challenge dose axis (logit response = 0). Infectivity was measured for the Iowa (○, —), Moredun (◊, — - - - —), TAMU (Δ, — — —), Maine (×, ------), and Glasgow (–,— - - —) isolates. Each data point represents the logit transformation of the proportional infectivity for two to seven replicate sets of cell cultures with 6 – 10 monolayers per replicate. The graph represents 100 datapoints derived from 805 individual wells of cell culture.

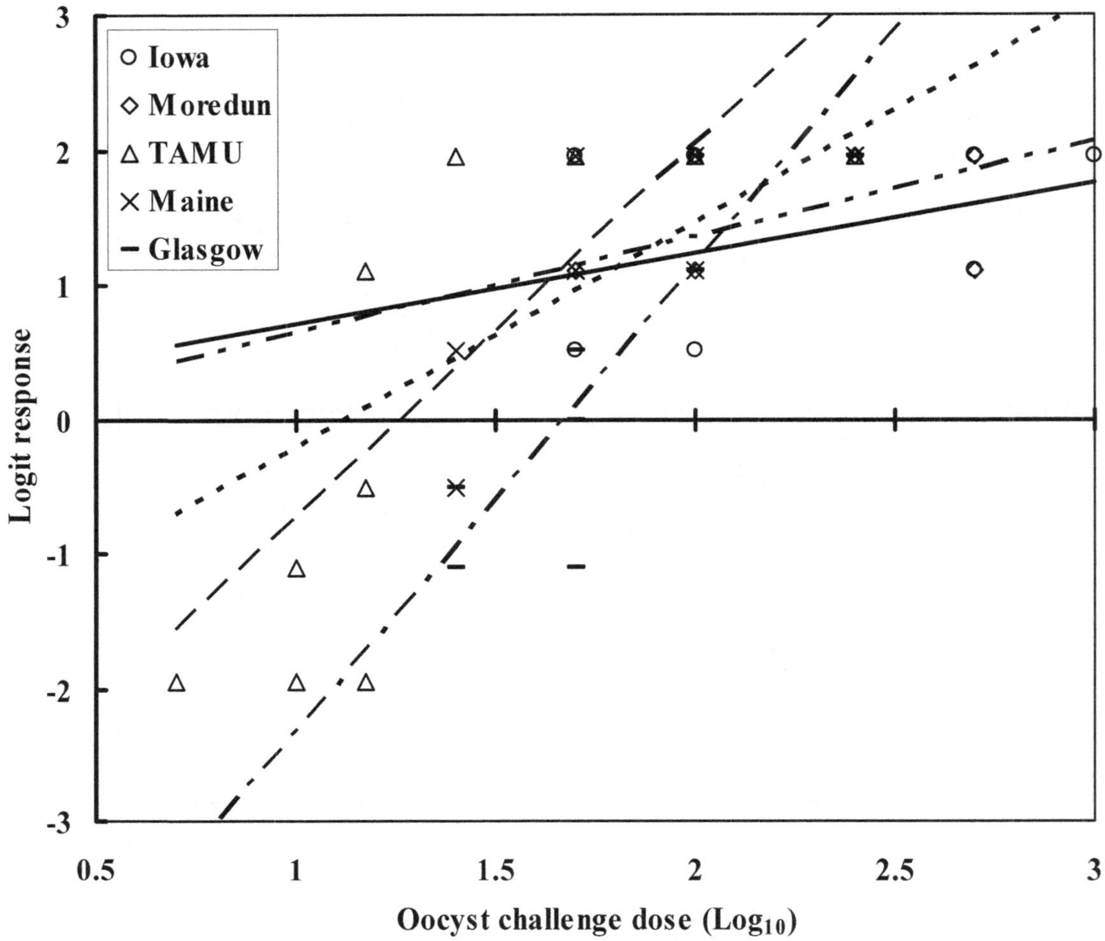

Figure 4.4 Dose response curves for five isolates of *C. parvum* in MDCK cells. Infectivity was assessed using an IFA assay. The ID_{50} is the point at which the regression line crosses the oocyst challenge dose axis (logit response = 0). Infectivity was measured for the Iowa (○, —), Moredun (◊, — — — —), TAMU (Δ, — — —), Maine (×, ------), and Glasgow (–,— — —) isolates. Each data point represents the logit transformation of the proportional infectivity for two to seven replicate sets of cell cultures with 6 – 10 monolayers per replicate. The graph represents 113 datapoints derived from 1,005 individual wells of cell culture.

Table 4.1
ID_{50} values for five isolates of *C. parvum* in CD-1 mice

Isolate	Oocyst doses	ID_{50} (average)*	ID_{50} (individual)†	95% CI‡	N (cages)§	N (total)**
Iowa	50 – 1,000	220	347	128 – 942	18	271
Moredun	25 – 500	28	16	9 – 25	25	361
Maine	25 – 250	22	21	10 – 40	21	264
TAMU	5 – 250	25	23	17 – 32	15	423
Glasgow	25 – 250	27	21	13 – 34	20	232

* Dose response curves were constructed by first averaging the proportional infectivity at each oocyst challenge dose and then calculating the response logit. ID_{50}s were calculated by solving the logistic regression for a logit response of zero.

† Dose response curves were constructed using individual proportional infectivity data converted to logit responses. ID_{50}s were calculated by solving the logistic regression for a logit response of zero.

‡ 95% confidence interval of the ID_{50}.

§ Number of cages of mice inoculated.

** Total number of mice inoculated.

Table 4.2
ID_{50} values for five isolates of *C. parvum* in HCT-8 cells

Isolate	Oocyst doses	ID_{50} (average)*	ID_{50} (individual)†	95% CI‡	N (sets)§	N (total)**
Iowa	50 – 1,000	77	106	52 – 154	13	144
Moredun	25 – 500	26	30	18 – 38	21	232
Maine	25 – 250	84	86	66 – 114	22	240
TAMU	5 – 250	34	27	21 – 34	11	237
Glasgow	25 – 250	81	81	60 – 110	22	192

* Dose response curves were constructed by first averaging the proportional infectivity at each oocyst challenge dose and then calculating the response logit. ID_{50}s were calculated by solving the logistic regression for a logit response of zero.

† Dose response curves were constructed using individual proportional infectivity data converted to logit responses. ID_{50}s were calculated by solving the logistic regression for a logit response of zero.

‡ 95% confidence interval of the ID_{50}.

§ Number of sets of cell monolayers inoculated.

** Total number of cell monolayers inoculated.

Table 4.3
ID_{50} values for four isolates of *C. parvum* in Caco-2 cells

Isolate	Oocyst doses	ID_{50} (average)*	ID_{50} (individual)†	95% CI‡	N (sets)§	N (total)**
Iowa	50 – 1,000	327	629	237 – 1,699	12	124
Moredun	25 – 500	40	45	28 – 70	22	204
Maine	25 – 250	104	118	89 – 156	20	240
TAMU	5 – 250	58	31	17 – 33	11	237
Glasgow	25 – 250	nd	nd	nd	nd	nd

See Table 4.2 for footnote explanation; nd, Not determined.

Table 4.4
ID_{50} values for five isolates of *C. parvum* in MDCK cells

Isolate	Oocyst doses	ID_{50} (average)*	ID_{50} (individual)†	95% CI‡	N (sets)§	N (total)**
Iowa	50 – 1,000	24	0.5††	0.02 – 13	14	175
Moredun	25 – 500	5	1.2	0.6 – 2	16	144
Maine	25 – 250	20	13	2 – 78	11	144
TAMU	5 – 250	17	18	6 – 53	11	230
Glasgow	25 – 250	3	1.5	0.4 – 7	11	120

See Table 4.2 for footnote explanation.
†† Iowa oocysts were incubated in 0.75% sodium taurocholate prior to inoculation of MDCK cells.

The ID_{50} for the Iowa isolate in CD-1 mice was 347 when calculated using individual replicate data and 220 when replicate percent infectivity values were averaged prior to logit calculations (Table 4.1). These values are considerably higher than those typically obtained for this isolate and strain of mouse at the University of Arizona. The ID_{50} for the Iowa isolate in CD-1 mice has previously been reported to be in the range of 60 – 87 oocysts (Finch et al., 1993; Korich et al., 2000a). Aggregate data obtained over a 3 year period from 22 separate analyses with oocysts of the Iowa isolate propagated in 10 individual calves, and utilizing over 1,700 mice, demonstrated that the average ID_{50} for this isolate in CD-1 mice was 100 with a 95% confidence interval of 84 – 119 (Korich et al., 2000b). However, the values of 347 and 220 obtained in the present investigation fall within the 80% prediction limits (26 – 390) of the earlier study (Figure 4.5) and should therefore be considered as valid values for this particular lot of oocysts in this individual experiment. This variation in ID_{50} values reflects the variability inherent in infectivity assays even when using the "gold standard" animal model. Various other mouse models have been used for measuring the infectivity of a variety of isolates of *C. parvum* with ID_{50} values in the range 60 – 1,000 (Lindsay, 1997).

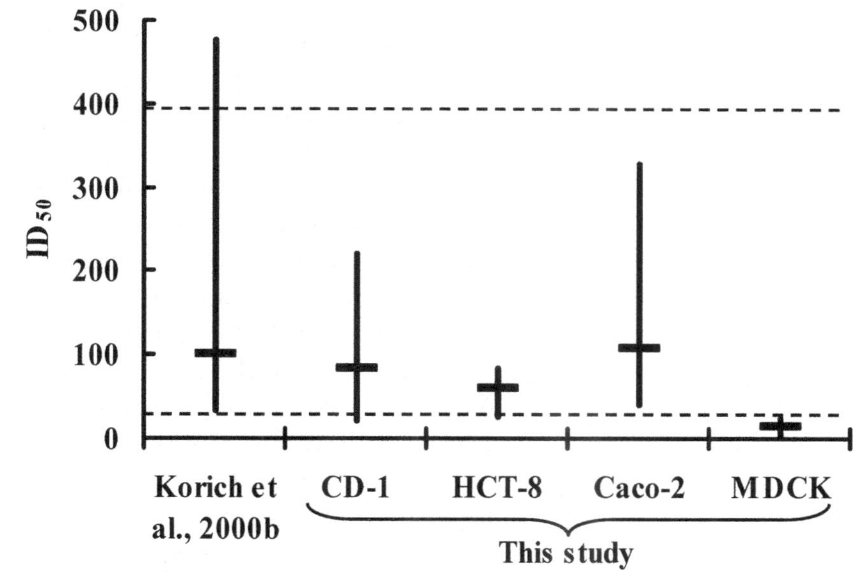

Figure 4.5. Average ID_{50} values (horizontal bars) and ranges (vertical bars) for the five isolates of *C. parvum* used in this study with each infectivity model compared to previously published data derived from 22 separate dose response determinations for just the Iowa isolate (Korich et al., 2000b). The horizontal dashed lines indicate the 80% prediction interval of the Korich data for the Iowa isolate.

The ID_{50} values calculated from averaged percent infectivity data for the five *C. parvum* isolates in this study ranged from 3 to 327 for the various cell and mouse models. The HCT-8/RT-PCR assay generated the narrowest span of ID_{50} values across all isolates with a 3.9-fold difference between the highest (Iowa) and lowest ID_{50} (TAMU), compared to 12-fold, 22-fold, and 20-fold differences for the MDCK/IFA, CD-1 mice, and Caco-2/RT-PCR assays, respectively. The mean and range of the ID_{50} values generated by the CD-1 mouse, HCT-8/RT-PCR cell culture, and Caco-2/RT-PCR cell culture assays, for all five isolates of *C. parvum* combined, were within the 80% prediction limits of the Iowa isolate ID_{50} (Figure 4.5), which were based on aggregate data from 22 individual dose response determinations (Korich et al., 2000b). The mean ID_{50} for all isolates combined as measured using the MDCK/IFA assay was less than the lower 80% prediction limit for the Iowa isolate. During the period covered by this project, infectivity was also measured for the KSU-1 isolate of *C. parvum* (Rochelle et al., 2004). With oocyst challenge doses of 36 – 818, the ID_{50} for the KSU-1 isolate in HCT-8 cells was 111 oocysts. The wide variation in calculated ID_{50} values for each isolate between the various assays makes it difficult to assign significance to the differences in ID_{50} between isolates. This variation is also reflected when comparing data from this study with previously published data for the few isolates that have been studied (Table 4.5). The figures may represent genuine differences in the levels of infectivity between different isolates or they may simply be a reflection of combined sources of variation in the assays. Consequently, no firm conclusions can be drawn concerning the relative infectivity of different isolates. Nevertheless, both the TAMU and Moredun isolates appeared to be more infectious than the Iowa isolate in CD-1 mouse, HCT-8, and Caco-2 cell culture assays with an average ID_{50} across all assays of 25 for TAMU compared to 355 for the Iowa isolate, a 14.2-fold difference. Although studies have been limited by the difficulties in obtaining sufficient suitable volunteers, *C. parvum* infection has also been studied in humans. The ID_{50} of the Iowa isolate in human volunteers has been reported as 75 – 132 oocysts (Dupont et al., 1995; Okhuysen et al., 1999) compared to nine oocysts for the TAMU isolate when presumed infection was used as the assay endpoint (Okhuysen et al., 1999), a 14.6-fold difference. However, when using actual oocyst excretion as the measure of infectivity, the ID_{50} of the TAMU isolate in humans was 125, higher than that of the Iowa isolate (75 oocysts) in the same study (Okhuysen et al., 1999). So, these comparisons in humans, mice, and cell culture indicate that the TAMU and Moredun isolates may be more infectious than the Iowa isolate.

There was considerable variation between replicates in the level of infection detected at each oocyst dose by all infectivity models. A similar level of variability was displayed for all assays; the average standard deviation in percent infectivity for all replicates of all isolates was 13.9%, 11.5%, 13.2%, and 10.7% for CD-1 mice, HCT-8 cells, Caco-2 cells, and MDCK cells, respectively. The range of standard deviations in percent infectivity between replicates for all isolates was 3.7 – 32.6 for CD-1 mice and 0 - 34.1 for the three cell culture assays combined. With an average standard deviation across all infectivity methods of 11.7%, standard power analysis demonstrated that with three experimental replicates (10 mice or 10 wells of cell culture per replicate) the minimum difference in infectivity that could be detected was 23.7% (Figure 4.6; Rosen et al., 2001). Reducing the detectable difference to 16% would require six replicates of 10 mice or 10 cell monolayers each (equivalent to a total of 60 mice or wells of cell culture) at each oocyst dose.

Infectivity assays are inherently variable due to the many factors involved in working with multiple living systems (oocysts, propagation animal, cell line or animal infection host). Natural lot-to-lot variability of the Iowa isolate was demonstrated by a review of 22 dose response studies in CD-1 mice over a 3-year period in which the ID_{50} values ranged from 33 to 476 (Korich et al., 2000b). Possible sources of such variation for *C. parvum* infectivity assays include:

- Inter-isolate variation due to genuine biological differences.
- Intra-isolate variation due to differences in oocyst production and handling procedures:
    - Oocyst age (post-shedding period).
    - Oocyst purification and storage procedures.
- Isolates containing two or more distinct strains/genotypes with genuine biological differences.
- Differences in the health and general condition of oocyst propagation animals.
- Inter-experimental variation in the susceptibility of the host (either cell monolayers or mice) to infection.
- Strain and age of mice or cells.
- Procedures used to enumerate oocyst doses prior to cell inoculation.

Table 4.5
Comparison of ID_{50} values

Assay model	Reference	ID_{50} for the following isolates: Iowa	TAMU	Moredun
CD-1 mice	This study	347 (128 – 942)*	23 (17 – 32)	16 (9 – 25)
HCT-8 cells	This study	106 (52 – 154)	27 (21 – 34)	30 (18 – 38)
CD-1 mice	Finch, 1993	60	nd	nd
CD-1 mice	Korich et al., 2000a	87	nd	nd
CD-1 mice	Korich et al., 2000b	100 (84 – 119)	nd	nd
Humans	Dupont et al., 1995	132	nd	nd
Humans	Okhuysen et al., 1999	75 (34 – 114)	125 (77 – 179)	nd
Humans	Okhuysen et al., 2002	nd	nd	300 – 375

* 95% confidence interval in parentheses.
nd, Not determined.

When using just the lower oocyst challenge doses of 5 – 25 oocysts for the TAMU isolate the ID_{50} values ranged from 18 in MDCK cells to 31 in Caco-2 cells and there was no significant difference between the dose response curves from HCT-8 cells and CD-1 mice ($P = 0.18$). However, using the entire TAMU isolate dataset with oocyst challenge doses of 5 – 250 to generate dose response curves, resulted in ID_{50} values of 19 in CD-1 mice (N = 25; 95% C.I., 15 – 23), 67 in HCT-8 cells (N = 27; 95% C.I., 18 – 234), and 73 in Caco-2 cells (N = 27; 95% C.I., 25 – 216) and there were significant differences in the slopes of the dose response curves between HCT-8 cells and CD-1 mice ($P<0.01$).

When data from all isolates were combined, the cell culture system that generated infectivity results most similar to the CD-1 mouse assay was HCT-8/RT-PCR ($r = 0.88$; Figure 4.7). More accurate representation of the human intestine may be an important factor in selecting a particular cell line for an infectivity assay, depending on the intended application of the assay. Cell cultures that provide the closest mimic of the human intestine may produce results that are the most accurate predictors of the potential for infection in humans. For example, Caco-2 cells spontaneously differentiate once they reach maximum confluence forming brush border microvilli and produce higher levels of brush border enzymes (alkaline phosphatase, sucrase-isomaltase and amino-peptidase) compared to non-differentiated cells (Pinto et al., 1983). This functional differentiation may make Caco-2 cells a better model of the human intestine than cell lines which do not possess these properties. HT29 cells can be induced to differentiate under glucose starvation conditions and differentiated HT29.74 cells (a clone of HT29) supported a five-fold increase in the development of *C. parvum* developmental stages compared to undifferentiated cells (Flanigan et al., 1991). However, the Caco-2/RT-PCR assay produced consistently lower measures of infectivity compared to CD-1 mice (Figure 4.8) while the MDCK/IFA method resulted in generally higher levels of infection (Figure 4.9). Although the relatively large variability in the data resulted in few significant differences between infectivity models, when differences did occur they consistently demonstrated lower levels of infection in Caco-2 cells compared to either of the other two cells lines or CD-1 mice.

A direct comparison between HCT-8 and MDCK cells using the same method to detect infections (RT-PCR) was only performed for the Glasgow isolate. The ID_{50} in HCT-8 cells was 81 (N=22; 95% C.I., 60 – 110) compared to 49 (N=27; 95% C.I., 35 – 67) in MDCK cells. Although the average ID_{50} determined by RT-PCR was lower in MDCK cells compared to HCT-8, the 95% confidence intervals of the ID_{50} values overlapped and there was no significant difference between the dose-response curves based on an analysis of covariance (P=0.48).

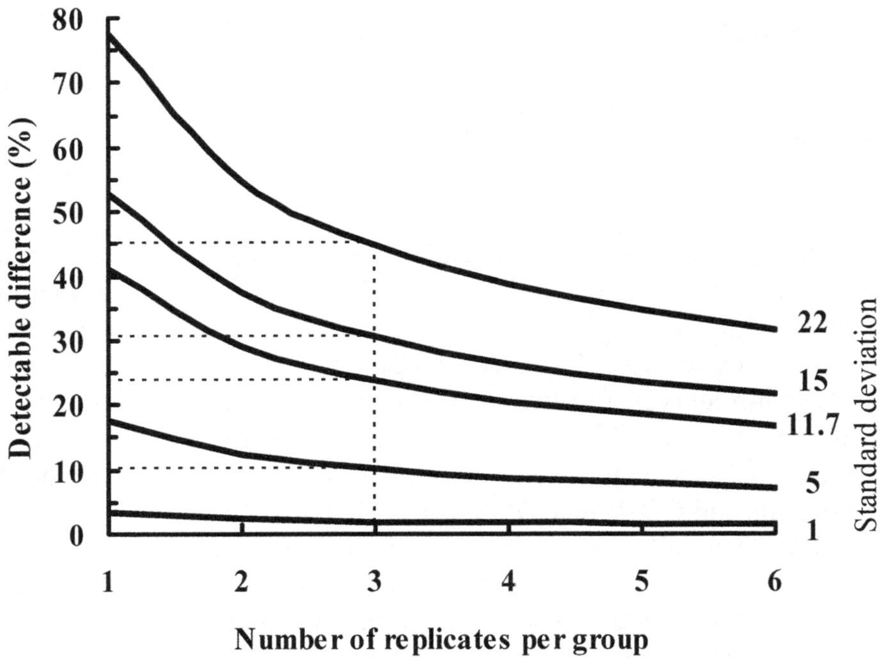

Figure 4.6 Power analysis of percent infectivity data for five isolates of *C. parvum* in all infectivity models using the indicated standard deviations. The average standard deviation across the entire dataset was 11.7%. The differences in percent infectivity that were detectable with three replicates are indicated by dashed lines. With 10 mice or wells of cell culture used to determine the percent infectivity, three replicates represents 30 mice or cell culture wells.

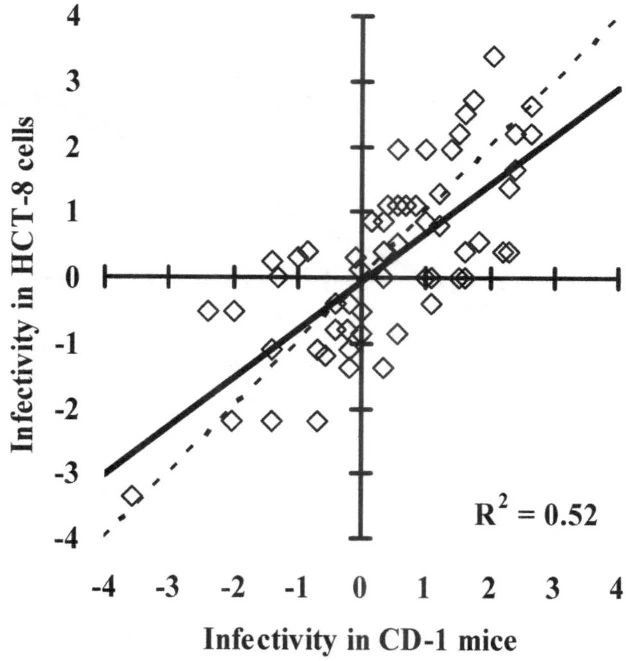

Figure 4.7 Comparison between HCT-8 cell culture and CD-1 mice for measuring infectivity of all five isolates combined. The dashed line indicates the ideal correlation. (Source: Rochelle et al., 2004)

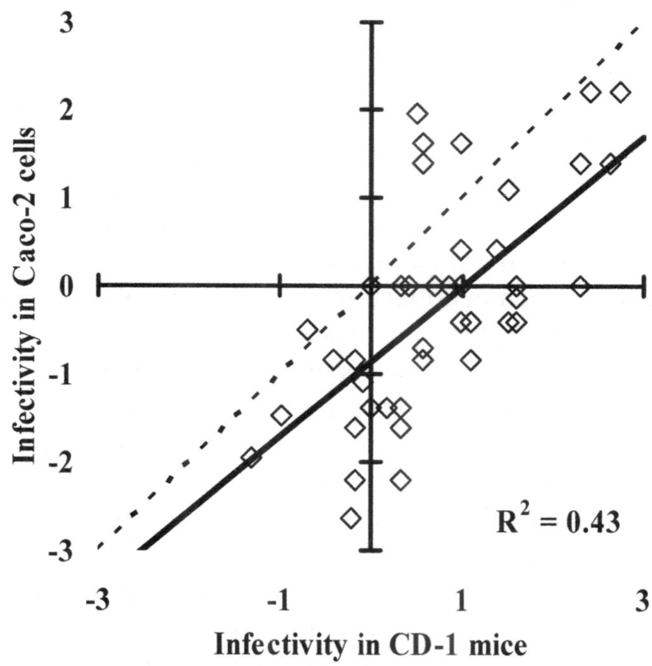

Figure 4.8 Comparison between Caco-2 cell culture and CD-1 mice for measuring infectivity of four isolates combined. The dashed line indicates the ideal correlation.

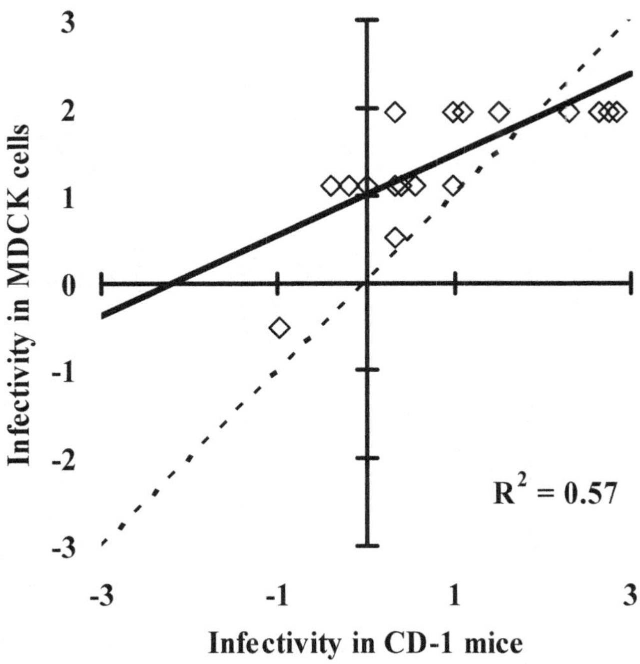

Figure 4.9 Comparison between MDCK cell culture and CD-1 mice for measuring infectivity of all five isolates combined. The dashed line indicates the ideal correlation.

COMPARISON OF CELL CULTURE AND MICE FOR MEASURING INFECTIVITY OF *CRYPTOSPORIDIUM HOMINIS*

A total of eight samples of oocysts recovered from human HIV patients attending a clinic in Lima, Peru were obtained. These oocysts were purified by discontinuous sucrose and cesium chloride centrifugation gradients and stored in 2.5% potassium dichromate. While this is not the ideal oocyst storage condition for infectivity studies, the project investigators had no control over the procedures used for oocyst collection, processing, or storage. These oocyst samples were received as a gift from another investigator and were not collected specifically for this project. Also, due to the logistics involved in collection and shipment of samples from foreign countries, the oocysts were approximately 60 days post-shedding at the time they were received.

Following genotyping of the isolates with the β-tubulin and COWP markers, three isolates were characterized as *C. hominis* (previously referred to as genotype 1 *C. parvum*), three were classified as *C. parvum* (previously genotype 2 *C. parvum*), and two of the samples failed to produce amplification products (see Chapter 3). Two *C. hominis* isolates were selected for further study and, in agreement with previous studies, neither isolate was able to infect either the propagation calves or CD-1 mice. Flow cytometry enumerated and sorted oocysts of one of the isolates (HIV1487) were used to construct a dose response curve in HCT-8 cells (Figure 4.10). The ID_{50} for this isolate was 741 oocysts in HCT-8 cell culture. However, for the reasons described above, this value cannot be compared directly with the ID_{50} values obtained for the five isolates of *C. parvum* because the latter isolates were all relatively fresh at the time of infection, had been preserved in saline antibiotic solution rather than potassium dichromate, and were freshly propagated in a production animal (bovine calves) instead of being harvested directly from critically ill human patients. A control dose comprising 100 oocysts of the Iowa isolate of *C. parvum* demonstrated 62.5% infectivity, demonstrating that the cells used for this particular experiment supported a level of infection consistent with expectations. The in-vitro excystation rate for these *C. hominis* oocysts was 81% compared to >90% for all of the various lots of *C. parvum* oocysts that were used throughout the project. Nevertheless, the results confirm that *C. hominis* can infect cell cultures but not standard animal infectivity models. The second *C. hominis* isolate also infected HCT-8 cells but the ID_{50} was not determined. Oocysts of *C. hominis* have been propagated to a limited extent in gnotobiotic piglets and limited infection in calves has also been reported (Akiyoshi et al., 2002). However, this project and other investigators (Peng et al., 1997; Widmer et al., 2000a) have demonstrated the inability of mouse infectivity models, including GKO mice, to support infection of *C. hominis*. Although propagation has been reported in gnotobiotic piglets, there is no routine widely applicable method for propagating the high numbers of oocysts of *C. hominis* that are required for extensive testing of infectivity and the efficacy of disinfectants. However, once such a technique becomes available, cell culture-based assays can be used to assess dose response curves and the response to disinfectants.

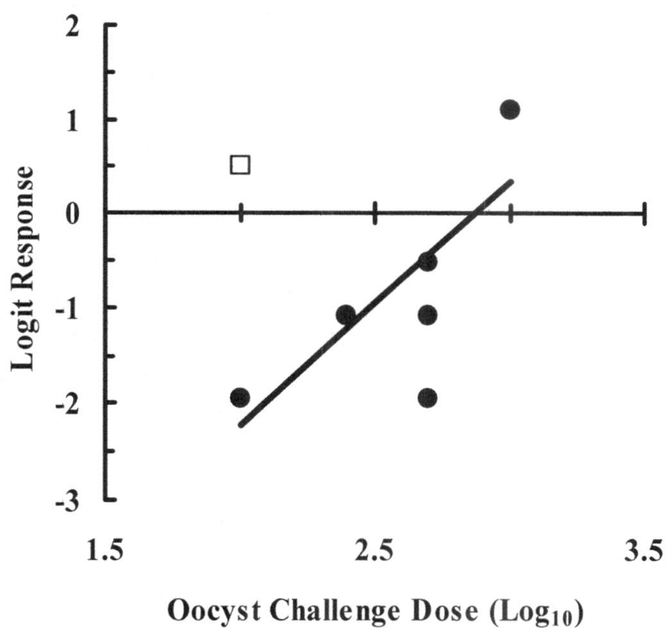

Figure 4.10 Dose response curve for an isolate of *C. hominis* in HCT-8 cell culture (•). Oocyst doses were prepared by flow cytometric sorting. A single dose of the Iowa isolate of *C. parvum* was also inoculated onto HCT-8 cells as an infection control (□).

SENSITIVITY OF CELL CULTURE ASSAYS

As stated above, it was not the intent of this study to maximize the sensitivity of infection in cell culture and then compare these assays with the CD-1 mouse assay. However, the sensitivity of cell culture assays for *C. parvum* infection can be increased by treatment of oocysts in a variety ways prior to inoculation of cell monolayers. The most commonly used treatments include incubation in 10% bleach or bile salts. Incubation in 10% bleach for 10 minutes is a routine procedure for simultaneous surface sterilization of oocysts (sometimes necessary if oocyst preparations are contaminated with high numbers of bacteria or yeast; see Table 3.2) and maximizing the sensitivity of infection in cell cultures (Upton 1997). Since the use of bleach as a pretreatment is not compatible with the use of oocysts for measuring the efficacy of disinfectants, incubation of oocysts in 0.75% sodium taurocholate prior to inoculation of cell monolayers was evaluated. These data were obtained from a number of experiments in which infectivity was compared with and without sodium taurocholate treatment of the oocysts and aggregated into a single figure (Figure 4.11). In contrast to a mean ID_{50} of 138 for untreated oocysts of the Iowa isolate in HCT-8 cells, the ID_{50} for oocysts following incubation in sodium taurocholate was 33. In some individual experiments the ID_{50} for sodium taurocholate treated oocysts was reduced by a factor of >10 compared to untreated oocysts. Aggregate data obtained during phase 1 of the project, comparing the three cell culture-based assays with mouse infectivity demonstrated that fresh oocysts of the Iowa isolate had an ID_{50} of 77 in HCT-8 cells (Table 4.2), 327 in Caco-2 cells (Table 4.3), and 0.5 in MDCK cells (Table 4.4). While the oocysts that were used to infect HCT-8 and Caco-2 cells were untreated, the oocysts used for MDCK infection were incubated in 0.75% sodium taurocholate prior to inoculation. Therefore, incubation in sodium taurocholate clearly increases the apparent infectivity of *C. parvum* in cell culture. However, the effect of such treatment on environmentally stressed oocysts or oocysts that have been exposed to disinfectants needs to be evaluated.

Following inoculation with oocysts, cell culture incubation periods typically vary from 24 h to 72 h. The longer incubation periods should permit development of sexual stages (see Figure 1.1) leading to in vitro amplification of the parasite and increased sensitivity. Using the CISH assay described in Chapter 3, Metropolitan has previously demonstrated that infectious foci contained 4.2 ± 2.2 parasite stages (mean ± standard deviation, N = 106) after post-inoculation incubation of HCT-8 cells for 24 h compared to 25.7 ± 8.9 stages (N = 63) following 48 h incubation (Rochelle et al., 2001b). A study using IFA as the detection method for infections in cell culture reported up to a 27-fold increase in the total number of parasite stages after 48 h incubation compared to 24 h (Slifko et al., 1997). For the current project, the sensitivity of the HCT-8/RT-PCR assay was evaluated using three replicate sets of eight wells for each oocyst challenge dose (24 cell monolayers per oocyst dose, oocyst dose = 5 – 25 oocysts per well). The ID_{50} for the TAMU isolate was 29.6 oocysts following 24 h incubation of inoculated cell monolayers compared to 28.0 after 48 h incubation. These ID_{50} values are in agreement with the ID_{50} for this isolate generated during an earlier part of the project (ID_{50} = 27, Table 4.2). Based on a two-factor ANOVA, there was no significant difference in the dose responses following differing incubation periods (P = 0.25). There was also no significant difference between 48 h and 72 h incubation for the Iowa isolate in HCT-8 cells (P = 0.84; 48 h ID_{50} = 126, 72 h ID_{50} = 129).

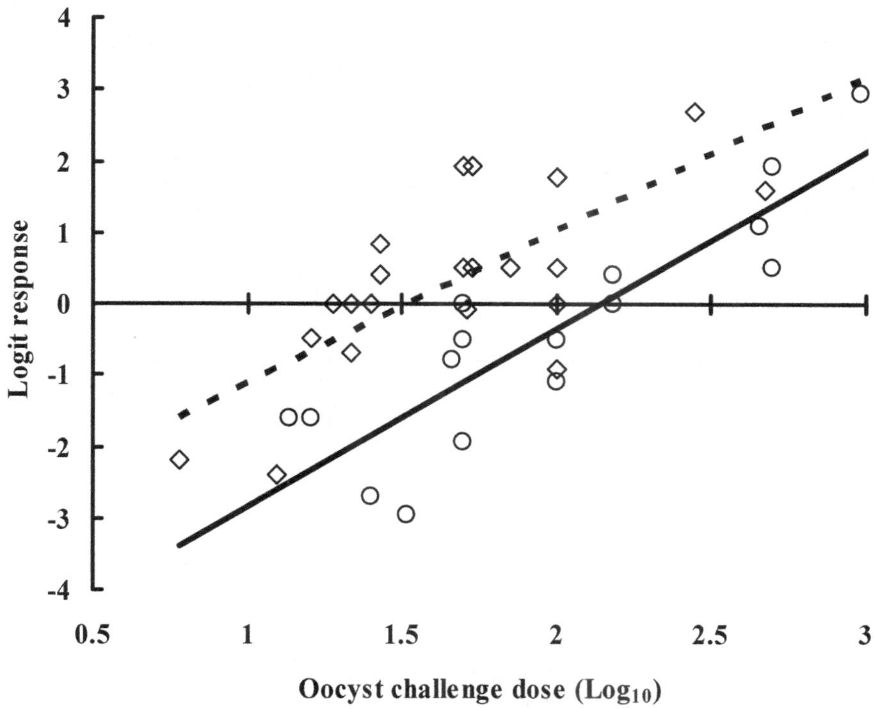

Figure 4.11 Effect of sodium taurocholate incubation on infectivity of the Iowa isolate in HCT-8 cell culture. Dose response curves were generated for oocysts that were inoculated directly onto cell monolayers (○, —; ID_{50} = 138) and oocysts that were incubated in 0.75% sodium taurocholate for up to 30 min prior to cell inoculation (◊, ----; ID_{50} = 33).

COMPARISON OF CELL CULTURE AND MICE FOR MEASURING INFECTIVITY OF OOCYSTS EXPOSED TO ENVIRONMENTAL WATER

A limited study was initiated to compare HCT-8 cell culture and the CD-1 mouse assay for measuring infectivity of oocysts that were exposed to untreated source water at ambient temperatures for prolonged periods of time. This part of the project was initially envisaged as being more extensive with a variety of different source water matrices tested and much longer exposure periods. However, the logistics of other aspects of the project limited this evaluation.

The ID_{50} of fresh oocysts of the Iowa isolate in HCT-8 cells was 117.5 at the start of the study, which was within the 95% confidence interval for this isolate in HCT-8 cells obtained in the earlier large comparison (see Table 4.2). Therefore, the oocysts used for this evaluation performed as expected in HCT-8 cells at the beginning of the study. Characteristics of this untreated source water were: turbidity = 0.67 NTU; total dissolved solids = 314 mg/L; total suspended solids = 0.6 mg/L; pH 8.4; conductivity = 569 µS/cm; total hardness = 118 mg/L; and dissolved oxygen = 10.9 mg/L. The temperature of the water during the 35 day exposure period ranged from 12°C to 19°C. The results demonstrated that there was no difference between CD-1 mice and HCT-8 cell culture for measuring the infectivity of oocysts stored in an environmental water sample (Table 4.6).

Table 4.6
Comparison of HCT-8 cells and CD-1 mice for assessing infectivity of oocysts exposed to environmental water for 35 days

	ID_{50} measured in the following models:	
Oocyst treatment	HCT-8 cells	CD-1 mice
Environmental water, 35 days	61.8	73.5
Environmental water, 35 days, bleach treated	24.2	–*
Control (4°C), 35 days	65.2	161.8

* Could not be calculated because oocyst challenge doses of 19 and 94 oocysts resulted in 5% of mice becoming infected while a challenge dose of 64 oocysts resulted in 0% infection.

IMPROVEMENTS AND MODIFICATIONS TO CELL CULTURE METHODS

Metropolitan's original cell culture assay for assessing infectivity of *C. parvum* involved growing HCT-8 cells in 1-well or 4-well chamber slides (Rochelle et al., 1997). Prior to starting this project the method was adapted to use 24-well culture plates and during the project period it was further adapted to a 96-well microtiter plate format. This format was used for cell culture, RNA extraction, and the RT-PCR assay; the benefits included reduced volumes for cell culture media, less processing time, more streamlined sample processing, and the use of multi-channel pipettors that reduce pipetting error and operator fatigue.

The 24-well and 96-well plate procedures were compared with 0.5% sodium hypochlorite-treated oocysts (Iowa isolate, 10% bleach for 10 min on ice followed by two washes in PBS). Inoculation with 50 oocysts resulted in 100% infectivity with both methods demonstrating that the 96-well plate method developed during this project had equivalent sensitivity to earlier procedures.

Many of the published cell culture infectivity assays for *C. parvum* included a washing step at 2 – 4 h post-inoculation to remove dead, empty, and unexcysted oocysts that attach to the cell monolayer but do not release sporozoites to initiate infection. Work at Metropolitan has previously demonstrated that this washing procedure may reduce the assay sensitivity because infectious oocysts that have not yet released their sporozoites are also removed (Rochelle et al., 1997). This aspect of the assay was evaluated again for the current project. Four replicate sets of eight wells each were inoculated with 200 oocysts and incubated at 35°C in a 5% CO_2 atmosphere. The culture medium was removed from two sets of wells after 2 h incubation and replaced with fresh growth medium; the monolayers were then incubated for the remainder of a 48 h period. Infectivity in the washed monolayers was 75% with 100 oocysts compared to 87.5% for monolayers that did not have inoculum oocysts removed. Since this difference is within the variability of the assay (Figure 4.6), post-inoculation washing of monolayers did not significantly reduce assay sensitivity. However, this aspect of *C. parvum* cell culture assays requires further evaluation.

Metropolitan has previously observed that consecutive subculturing of Caco-2 cells in collagen-treated vessels reduced their susceptibility to infection by *C. parvum* and this phenomenon was exploited to develop a Caco-2 subclone that was resistant to infection, even with 1×10^5 oocysts (De Leon and Rochelle, unpublished observations). We have also observed that higher passage numbers of Caco-2 and HCT-8 cells become less susceptible to infection regardless of the type of culture vessel. In a direct comparison using HCT-8 cells grown in 24-well plates and inoculated with 100 oocysts (Iowa isolate) purified by immunomagnetic separation, the infectivity was 80% in monolayers that had been through 11 passages compared to 60% for those that had been subcultured 40 times. Consequently, cell cultures are only used through passage 30 for measuring *C. parvum* infectivity. Different cell culture handling and maintenance procedures, including the use of differing cell line passage number, may be a source of variability between laboratories, leading to increased or decreased sensitivity in one laboratory compared to others. Although HCT-8 cells appear to have been adopted as the standard cell line for water industry-related applications of *C. parvum* infectivity assays, procedures for cell line maintenance and culture need to be standardized.

This project evaluated infectivity of *C. parvum* in three cell lines (HCT-8, Caco-2, and MDCK). Different cell lines require different media formulations for optimum growth but these optimum host growth conditions may not be conducive to maximum parasite development. Standard cell culture media contain all of the essential nutrients for cell growth and they are typically supplemented with FBS, antibiotics, and other additives. An optimized media formulation containing HEPES, glucose, glutamine, ascorbic acid, folic acid, insulin, 4-aminobenzoic acid, and calcium pantothenate was reported to yield a 10-fold increase in the number of *C. parvum* developmental stages compared to a standard medium (Upton 1997). In addition, there are conflicting reports on the optimum FBS concentration required once cells have been infected and it has been suggested that FBS (or at least certain formulations of FBS) may contain components that inhibit parasite growth (Arrowood et al., 1994). In the current study, cells were grown in RPMI-1640 (HCT-8) or Eagle's minimal essential medium (Caco-2) containing 10% FBS prior to infection and 2% FBS following inoculation with oocysts. MDCK cells were grown in serum-free medium. So, despite the general equivalency of cell culture with mouse infectivity demonstrated by this project, research is still needed to optimize and standardize media formulations to maximize infectivity and improve reproducibility.

Most of the cell culture assays for *C. parvum*, including those used in this project, utilize two-dimensional monolayers of host cells grown on microscope slides, cover-slips, permeable membranes, or in multi-well plates. However, it has been demonstrated that a low-shear, low turbulence, rotating-wall culture vessel promoted the development of three-dimensional cell aggregates that more accurately modeled human differentiated tissue for studying infection with *Salmonella enterica* (Nickerson et al., 2001). Up to 160 genes were expressed in *S. enterica* cultured under these conditions but were not expressed under standard culture conditions. A similar system maybe useful for studying in vitro *C. parvum* infection and could provide the missing component that currently limits in vitro oocyst propagation.

CHAPTER 5
RESULTS AND DISCUSSION

PHASE 2: COMPARISON OF CELL CULTURE AND MICE FOR ASSESSING DISINFECTION OF *CRYPTOSPORIDIUM PARVUM*

The majority of *C. parvum* inactivation studies have used mouse models to determine the levels of disinfection. Cell culture-based methods have been used in a more limited manner to evaluate chlorine, chlorine dioxide, and UV light disinfection of *C. parvum*. However, there have been few direct comparisons between cell culture and mouse assays. Phase 2 of this project involved using the CD-1 mouse assay and cell culture to assess the efficacy of UV and ozone for inactivating oocysts of *C. parvum*.

INACTIVATION OF *C. PARVUM* BY UV LIGHT

There was good agreement between the HCT-8 cell culture and CD-1 mouse assays for measuring UV inactivation of *C. parvum* (Figure 5.1) although in these particular experiments the mouse assay demonstrated greater inactivation (1.67-$\log_{10}$) at the lowest UV dosage used for these direct comparisons (2.2 mJ/cm^2) compared to the HCT-8 cell culture assay (0.39-$\log_{10}$). Forcing the regression lines through the origin generated indistinguishable inactivation curves for cells and mice. Combined data demonstrated that 2-$\log_{10}$ (99%) inactivation was achieved with a medium-pressure UV dosage of 5.6 mJ/cm^2. These results are in accordance with earlier studies using the same HCT-8 cell culture/RT-PCR methodology used in this project (Rochelle et al., 2004), a published study that used a less robust, less replicated version of the HCT-8/RT-PCR assay (Mofidi et al., 2001), and various published mouse-based studies of UV inactivation (Figure 5.2). The UV inactivation data generated for this project was obtained from experiments using the Iowa, Moredun, and TAMU isolates of *C. parvum*. Although only a limited amount of data was collected for each isolate, the results generally demonstrated that the different isolates responded to UV light in the same manner. In a previous study in which five isolates of *C. parvum* (Iowa, Moredun, Maine, TAMU, and KSU-1) were exposed to both low-pressure and medium-pressure UV light in a more extensive series of experiments using HCT-8 cell culture, there was no significant difference in the inactivation curves of the various isolates (Rochelle et al., 2004). A UV dosage of 6 mJ/cm^2 resulted in inactivation levels of 2.4-$\log_{10}$ for the Iowa isolate, 2.2-$\log_{10}$ for the Moredun isolate, 2.3-$\log_{10}$ for KSU-1, 1.8-$\log_{10}$ for the TAMU isolate, and 2.8-$\log_{10}$ for the Maine isolate.

These results demonstrate good overall congruence of mouse and cell culture assays for measuring UV disinfection of *C. parvum*. The HCT-8 cell culture/RT-PCR assay should therefore be considered equivalent to mouse infectivity for measuring UV inactivation of *Cryptosporidium* oocysts. At present, it is the only infectivity-based method available for measuring UV inactivation of *C. hominis* oocysts since this organism does not infect standard mouse models.

The overall agreement in UV inactivation data generated by various laboratories using different assay methods and a variety of isolates provides a strong measure of confidence to the conclusion that UV efficiently inactivates *C. parvum*. The UV dosage requirements needed to achieve various levels of inactivation, based on combined UV disinfection data from this project, two earlier projects using variations of the HCT-8/RT-PCR assay, and selected mouse-based studies, are displayed in Table 5.1. The required dosages based on the upper 95% confidence interval of the mean inactivation are in general agreement with the required dosages included in the recently proposed Long Term 2 Enhanced Surface Water Treatment Rule (LT2ESWTR, USEPA, 2003).

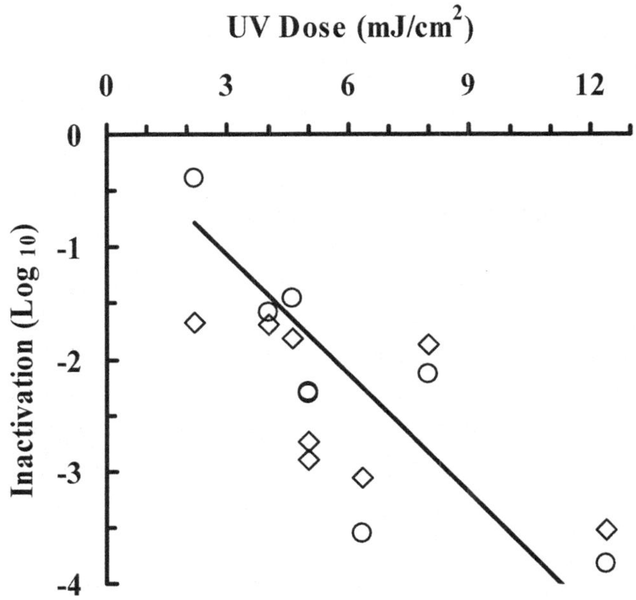

Figure 5.1 Comparison of the CD-1 mouse (◊) and HCT-8 cell culture/RT-PCR (○) assays for measuring inactivation of *C. parvum* oocysts by medium-pressure UV light. Regression lines forced through the origin were identical.

Table 5.1
UV dosage requirements for various levels of inactivation

	UV dosage for indicated inactivation (mJ/cm^2)			
Inactivation	Mean*	95% confidence*	90% prediction*	USEPA-LT2†
0.5-Log_{10}	1.37	1.85	4.08	1.6
1-Log_{10}	2.73	3.21	5.45	2.5
2-Log_{10}	5.57	5.95	8.18	5.8
3-Log_{10}	8.2	8.68	10.92	12

* Based on a single regression through all of the data displayed in Figure 5.2, including data generated by this project.

† Based on values included in the proposed Long Term 2 Enhanced Surface Water Treatment Rule (LT2ESWTR, USEPA, 2003).

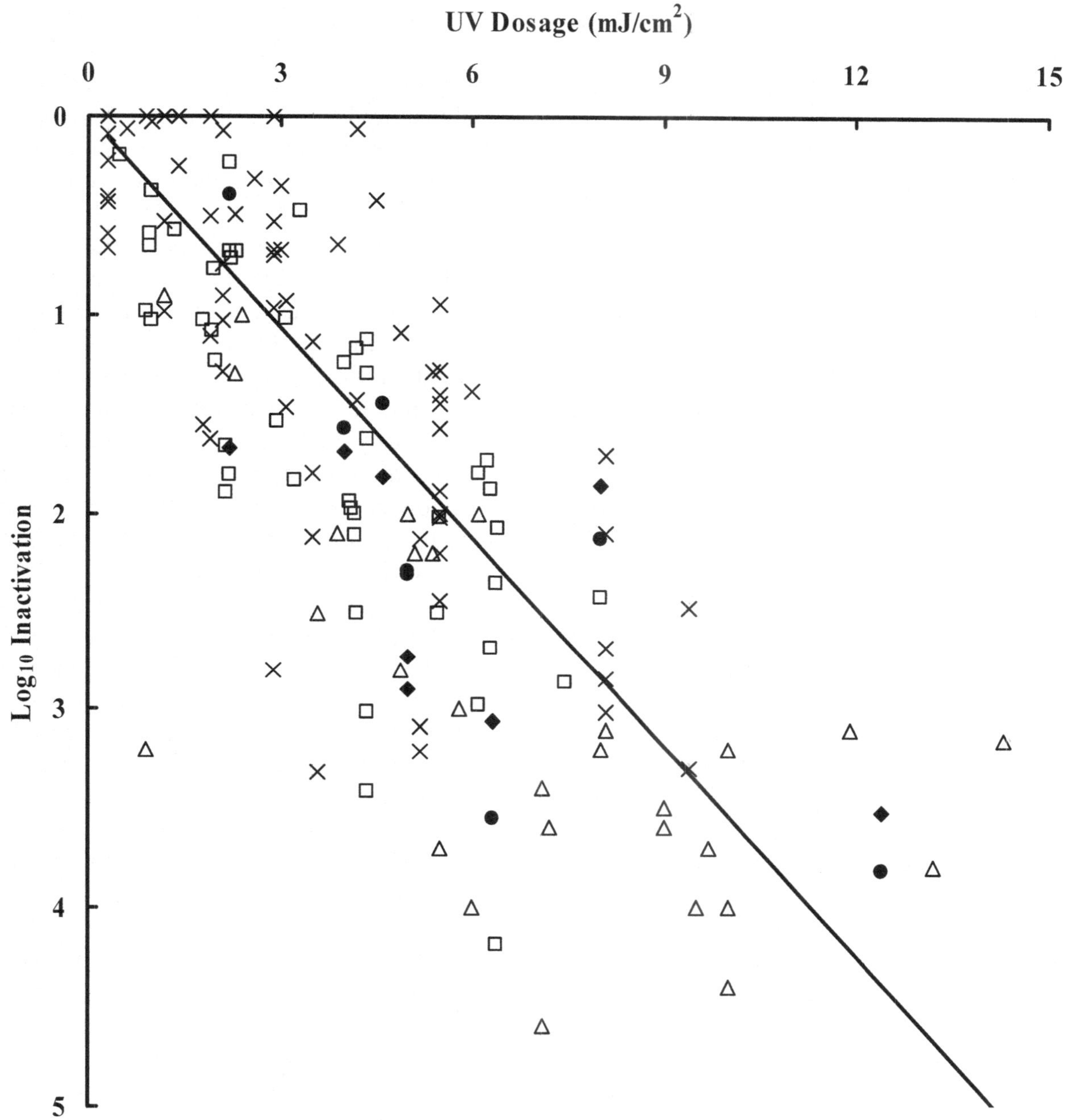

Figure 5.2 Summary of UV inactivation of *C. parvum*. Data were generated using HCT-8 cell culture (•) and CD-1 mice (♦) for this project, HCT-8 cell culture for another AwwaRF-funded project (□, Rochelle et al., 2004), an earlier, less robust version of the HCT-8 cell culture/RT-PCR assay (×, Mofidi et al., 2001), and a variety of published studies employing mouse infectivity models (Δ). The regression line was plotted through the HCT-8 cell culture/RT-PCR data generated by this project only.

OZONE REACTOR DESIGN CONSIDERATIONS

Ozone disinfection experiments were performed in three types of reactor: simple batch, semi-batch, and a single continuously stirred tank reactor (CSTR). Batch reactors are the simplest type of experimental disinfection vessel, and generally the most efficient in terms of the amount of inactivation observed. They are essentially closed systems in which a solution of dissolved ozone (typically in buffer or natural waters) is mixed with the test organisms. A sample of the organism suspension is removed prior to addition of ozone to obtain the initial viable organism concentration (N_0) and then at periodic intervals following addition of ozone to determine the surviving fraction of organisms (N). Neither the ozone nor organisms are replenished during the course of the test. Semi-batch reactors, referred to as ideal plug-flow reactors (PFR), have ozone continuously passing through the suspension matrix containing organisms but there is no liquid flow. The contents of the reactor are mixed so that the concentration of organisms and ozone is uniform throughout the liquid phase. CSTRs are a type of continuous flow reactor in which both the ozone is continuously bubbled through the reactor and the matrix liquid containing the organisms continuously flows through the reactor in the counter direction to ozone. In the CSTR configuration, the counter-flowing ozone and liquid phases are continuously stirred. While this type of reactor represents the best model of full-scale disinfection plants, it is the most inefficient systems in regards to the amount of inactivation observed. Figure 5.3 displays a hypothetical comparison of inactivation using an ideal CSTR and an ideal PFR assuming linear Chick-Watson disinfection kinetics. For each reactor type, high and low lethality constants (k) were used that bracketed the range of values measured in previously published ozone inactivation study (Oppenheimer et al., 2000). While comparable inactivation is expected at low CT values, at higher CT values the PFR will achieve substantially higher levels of inactivation. With a k_e value of 0.5 L/mg.min, which is representative of lethality constants at higher temperatures (>10°C), and at a CT of 15 mg.min/L, a PFR should achieve approximately 3.3-$\log_{10}$ inactivation compared to only 0.93-$\log_{10}$ inactivation in a CSTR. Moreover, to achieve a 3-$\log_{10}$ inactivation using a single CSTR and with a k_e value of 0.5 L/mg.min, a CT of 200 mg.min/L would be required, which would necessitate maintaining a dissolved ozone residual of 9 mg/L for a detention time of 4 hours.

At a lower k value, representative of lethality constants at low temperatures (~3°C), the magnitude of the difference between the two reactor types is substantially less. The PFR should achieve 1.3-$\log_{10}$ inactivation at a CT of 60 mg.min/L while the CSTR should achieve 0.6-$\log_{10}$ inactivation. A CT of 90 mg.min/L would be required to achieve 2-$\log_{10}$ inactivation in the PFR while the CSTR would require a CT of 2,100 mg.min/L. From a hydraulic perspective, full-scale ozone contactors can be described as multiple CSTRs arranged in series. Therefore, the potential disinfection capability of full-scale treatment plants will be higher than can be achieved using the single CSTR that was used for this project. Conversely, because full-scale water treatment plants do not operate as plug flow reactors, semi-batch disinfection experiments generally overestimate the level of inactivation that can be achieved at full-scale.

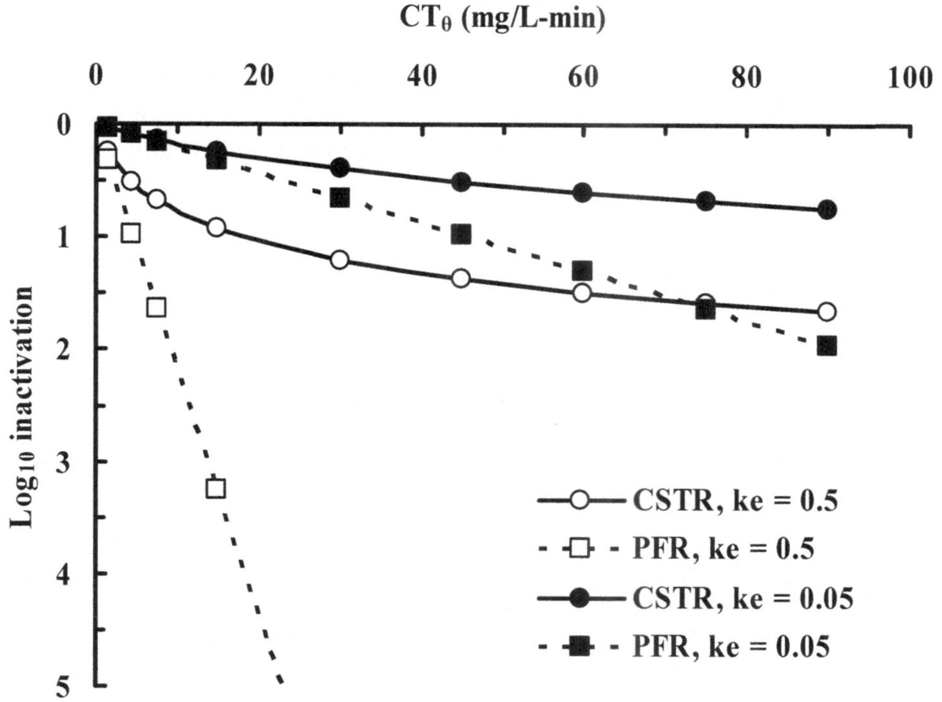

Figure 5.3 Comparison of theoretical disinfection in different types of reactor with different lethality constants (k_e). CSTR, Continuously stirred tank reactor; PFR, Plug flow reactor. In this example, T = 15 min for all values of CT.

OZONE DISINFECTION OF *BACILLUS SUBTILIS* SPORES

Ozone inactivation experiments were performed with *B. subtilis* spores, using the same 2 L semi-batch reactor that was used for the majority of the *C. parvum* experiments. The majority of the data was collected at 5°C with more limited data at 1°C and 15°C (Figure 5.4). Based on unforced linear regressions, CT values of 19.5, 34.4, and 59.5 mg.min/L were necessary to achieve 3-$\log_{10}$ inactivation at 15, 5, and 1°C, respectively. There is limited published information on ozone inactivation of *B. subtilis* spores at these temperatures for comparison. Previous work at MWD demonstrated 1-$\log_{10}$ inactivation with a CT of 7.7 mg.min/L at 10°C in a reactor comprising three CSTRs (Coffey et al., 2001). At 20°C and pH 7 in a semi-batch reactor, 2-$\log_{10}$ inactivation was achieved at CT = 7 mg.min/L (Larson and Marinas, 2003). The effect of pH on disinfection kinetics was demonstrated with 2-$\log_{10}$ inactivation at CT = 6.3 mg.min/L at pH 8.2 and CT = 3.2 mg.min/L at pH 6.3 in a batch reactor at 20 – 22°C (Facile et al., 2000). Similar results were also obtained for *Bacillus cereus* with a CT of 6 mg.min/L resulting in 2-$\log_{10}$ inactivation in a batch reactor at 22°C and pH 6 (Finch and Choe, 1999). The most extensive study reported a CT_{lag} of 10 mg.min/L at 5°C, during which little inactivation occurred, decreasing to 1.4 mg.min/L at 30°C (Driedger et al., 2001). At all temperatures tested the lag phase was followed by a linear decrease in spore viability. The CT required to achieve 3-$\log_{10}$ inactivation ranged from 3 mg.min/L at 30°C to 17.6 mg.min/L at

5°C. Finally, in a pilot-scale trial, an ozone CT of 19 mg.min/L was necessary to achieve 2-log_{10} inactivation of indigenous aerobic sporeforming bacteria (Owens et al., 2000). Although it is difficult to make comparisons across the various studies, these published results show good overall agreement with the expected level of inactivation based on the three temperatures used in this project. Therefore, this series of experiments clearly demonstrated that the semi-batch ozone reactor was operating as expected, delivering ozone to the target cells, and capable of achieving at least 6-log_{10} inactivation.

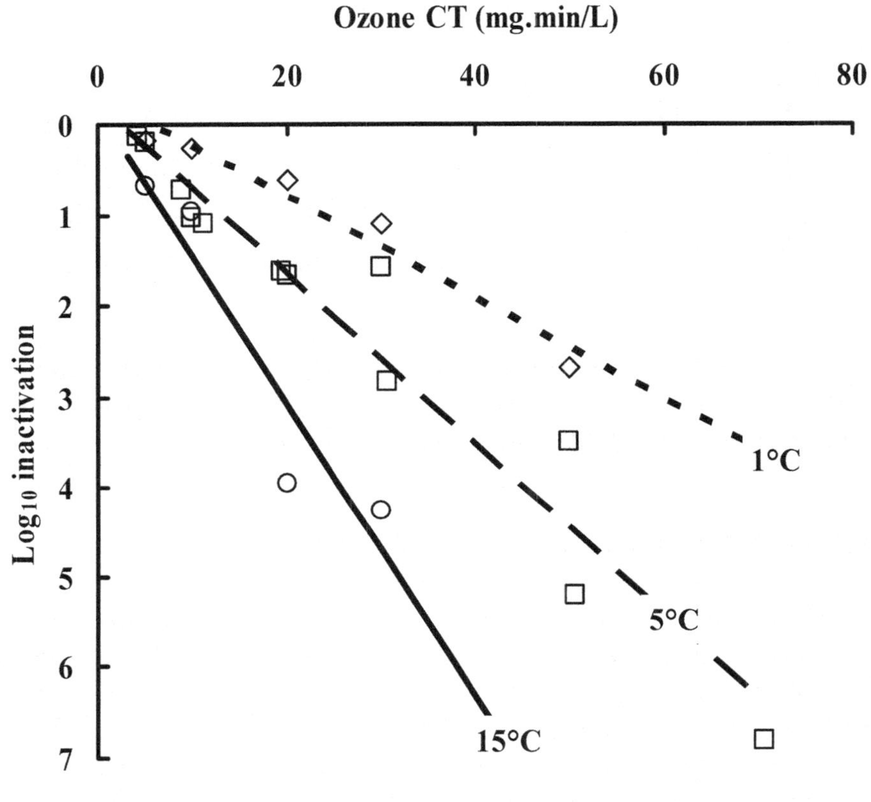

Figure 5.4 Inactivation of *Bacillus subtilis* spores by ozone in a semi-batch reactor at 15°C (○), 5°C (□), and 1°C (◊). Spores were suspended in filtered source water as described in Chapter 3.

OZONE INACTIVATION OF *CRYPTOSPORIDIUM*

Dose response curves were generated for control oocysts used in each ozone disinfection experiment. This ensured that inactivation was calculated based on a specific dose response relationship for the actual oocysts that were used in each trial. This approach should eliminate variability in disinfection data that results from comparing the response of multiple batches of treated oocysts to a dose response relationship generated by a single batch of oocysts. Dose response curves based on aggregate data for the untreated, control oocysts used in all of the ozone disinfection experiments demonstrated that these oocysts responded as expected in their respective infection hosts (Figure 5.5). The ID_{50} values in CD-1 mice, HCT-8 cells, and MDCK cells were 42, 65, and 40, respectively. These values are in general agreement with the ID_{50} values calculated in phase 1 of the project (see Tables 4.1 – 4.5). The ID_{50} in CD-1 mice was much lower than the calculated value derived from the initial dose response experiment with the Iowa isolate in phase 1 of the project (220 oocysts; Table 4.1), although it was closer to more typical ID_{50} values for this isolate. This provides a further demonstration of the inherent variability in infectivity assays. For HCT-8 cells, the ID_{50} for these control oocysts was similar to the earlier value in phase 1 (77 oocysts; Table 4.2).

Ozone Inactivation of *C. parvum* in a Continuously Stirred Tank Reactor (CSTR)

Initial ozone disinfection experiments used a 460 mL CSTR. Tracer tests were conducted to assess reactor performance by first establishing a known flow rate through the reactor and then by adding a tracer slug directly into the reactor. A 20 mL slug of 1 mM indigo solution was used as the tracer, which was a conservative measure as ozone was not flowing through the reactor. Indigo concentrations were measured by attaching a spectrophotometer with an automatic sampler to the reactor sample port and measuring absorbance at 600 nm in 30 – 60 second intervals. Figure 5.6 shows normalized tracer data measured for the bench-scale, single CSTR reactor used for this study. The profile for an ideal reactor (solid line) is plotted along with measured data. Two hydraulic detention times were tested to assess hydraulic performance at different flow rates. The theoretical ideal and measured data are in close agreement indicating that the bench-scale CSTR was behaving similarly to an ideal reactor. Theoretical t_{10}/t values for a single CSTR configuration are 0.1 and the measured t_{10}/t values were within 5% of the ideal values. The difference probably resulted from a combination of non-ideal behavior and experimental errors such as inaccuracy in flow measurement or tracer measurements.

Dose response data for the control oocysts used in the five CSTR experiments that were performed are displayed in Table 5.2. The calculated ID_{50} values ranged from 7.9 to 132.

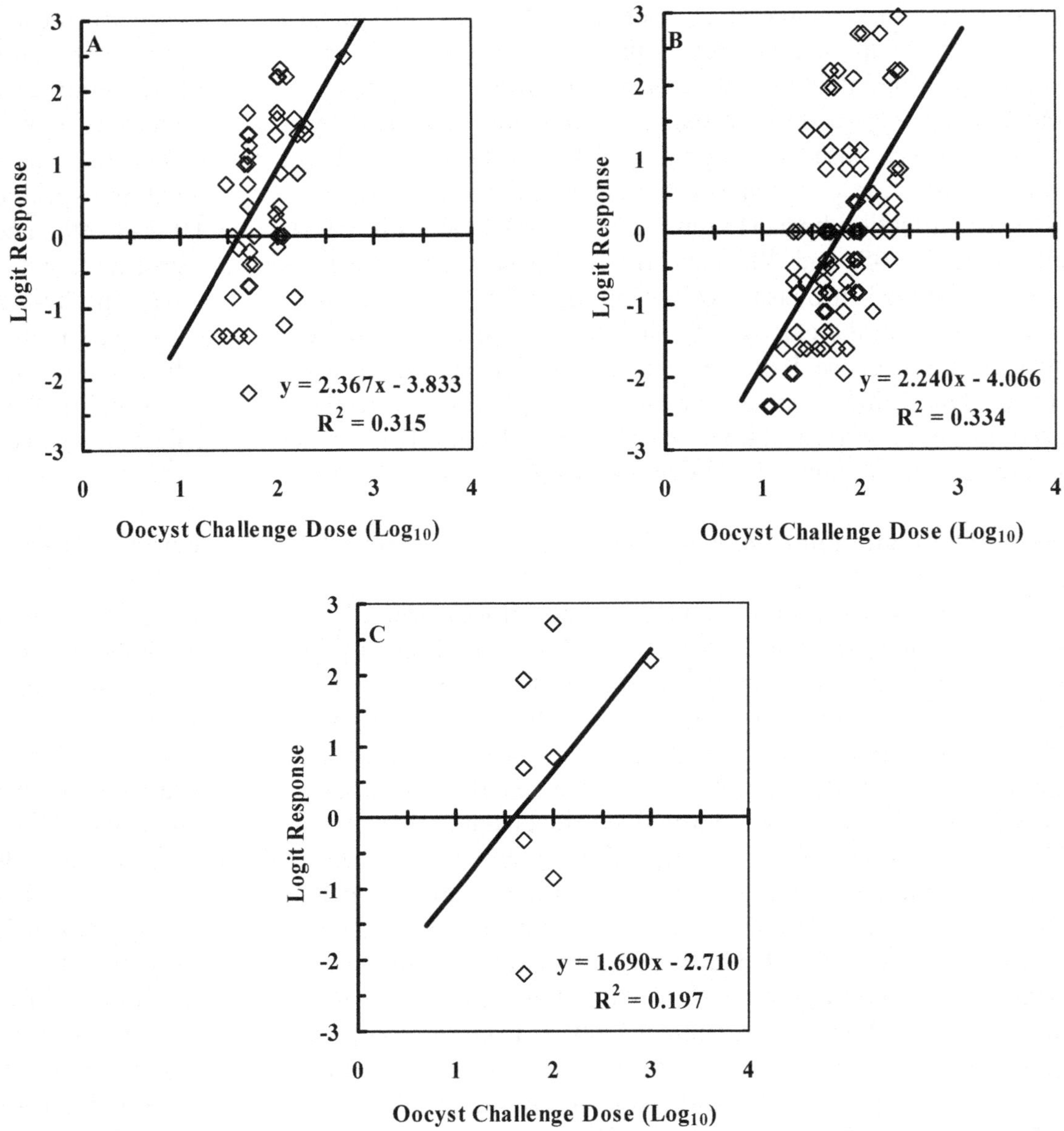

Figure 5.5 Aggregate dose response curves for control oocysts used in all ozone inactivation studies. Oocysts were not exposed to ozone and infectivity was measured in CD-1 mice (A), HCT-8 cell culture (B), and MDCK cell culture (C).

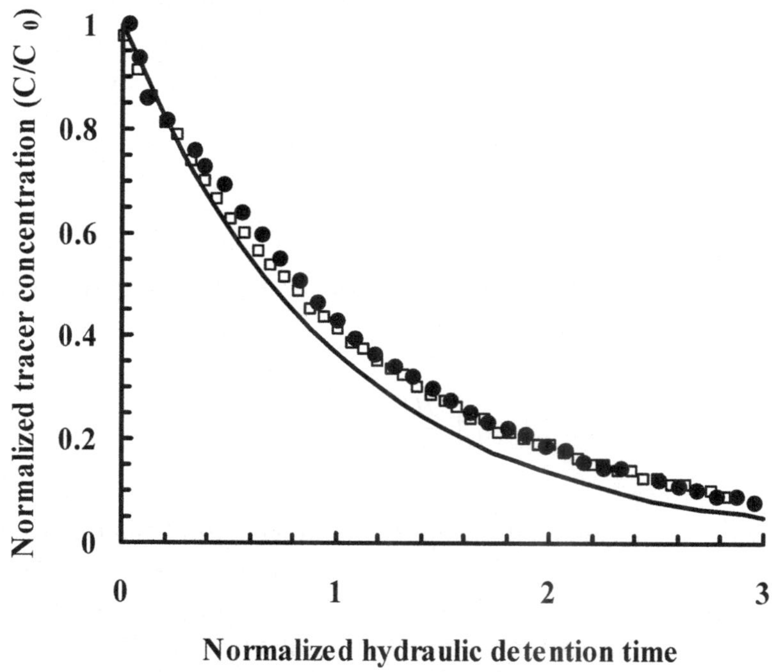

Figure 5.6 Continuously stirred tank reactor tracer test results showing measured data (symbols) and a theoretical ideal reactor (solid line). Data were obtained from a 460 mL reactor with hydraulic residence times of 5.6 min (•) and 15.8 min (□).

Table 5.2
Dose response data for control oocysts (CT = 0) used in continuously stirred tank reactor experiments

	HCT-8 Cells		CD-1 Mice	
Trial	Logit Dose Response	ID_{50}	Logit Dose Response	ID_{50}
1	LR = (1.799 × Log_{10} dose) – 3.421	79.8	LR = (5.060 × Log_{10} dose) – 8.495	47.7
2	LR = (3.179 × Log_{10} dose) – 6.096	82.7	LR = (1.195 × Log_{10} dose) – 1.070	7.9
3	LR = (4.576 × Log_{10} dose) – 7.624	46.3	LR = (2.803 × Log_{10} dose) – 5.186	70.7
4	LR = (0.486 × Log_{10} dose) – 0.932	82.6	LR = (3.790 × Log_{10} dose) – 5.817	34.3
5	LR = (2.805 × Log_{10} dose) – 5.948	132	nd	nd

Inactivation data for the CSTR experiments were calculated in two ways:
1) The proportional infectivity for each dose of ozone-treated oocysts was transformed by the logit function and then used to calculate the apparent number of viable oocysts in the sample (N) by solving the dose response model for each particular batch of oocysts (Table 5.2). The actual oocyst dose, enumerated by IFA well counts, was used as N_0 in the calculation of inactivation:

$$Inactivation = Log_{10}\left(\frac{N}{N_0}\right) \tag{5}$$

2) ID_{50} values were calculated for control and ozone-treated oocysts and inactivation was calculated using a modification of equation 4:

$$Inactivation = Log_{10}\left(\frac{ID_{50}^{ozone}}{ID_{50}^{control}}\right) \tag{6}$$

The inactivation data obtained using CD-1 mice conformed reasonably well to the theoretical expectations for a single CSTR modeled with $k_e = 0.5$ (Figure 5.3) which is consistent with the majority of the data being collected at 15°C (Figures 5.7A and 5.8). In the CSTR configuration a mean 2-$\log_{10}$ inactivation was achieved at a CT_{10} of 38.5 mg.min/L.

Up to this point in the project, the HCT-8 cell culture/RT-PCR assay demonstrated an overall good agreement with the CD-1 mouse assay for measuring the infectivity of fresh, untreated oocysts and oocysts exposed to UV light. Therefore, the initial CSTR-based disinfection data obtained using HCT-8 cells were unexpected. Although there was considerable scatter in the data (as there was for the CD-1 derived data also), the results indicated that oocysts exposed to doses of ozone less than 20 mg.min/L were predominantly more infectious than the non-exposed control oocysts indicating that oocysts were stimulated rather than inactivated (Figures 5.7B and 5.8). The highest level of inactivation obtained in the CSTR measured using HCT-8 cells was 1.14-$\log_{10}$.

MDCK cells were used to measure infectivity in some CSTR experiments but dose response data was not available for the control oocysts because all oocyst challenge doses resulted in either 0% or 100% infection.

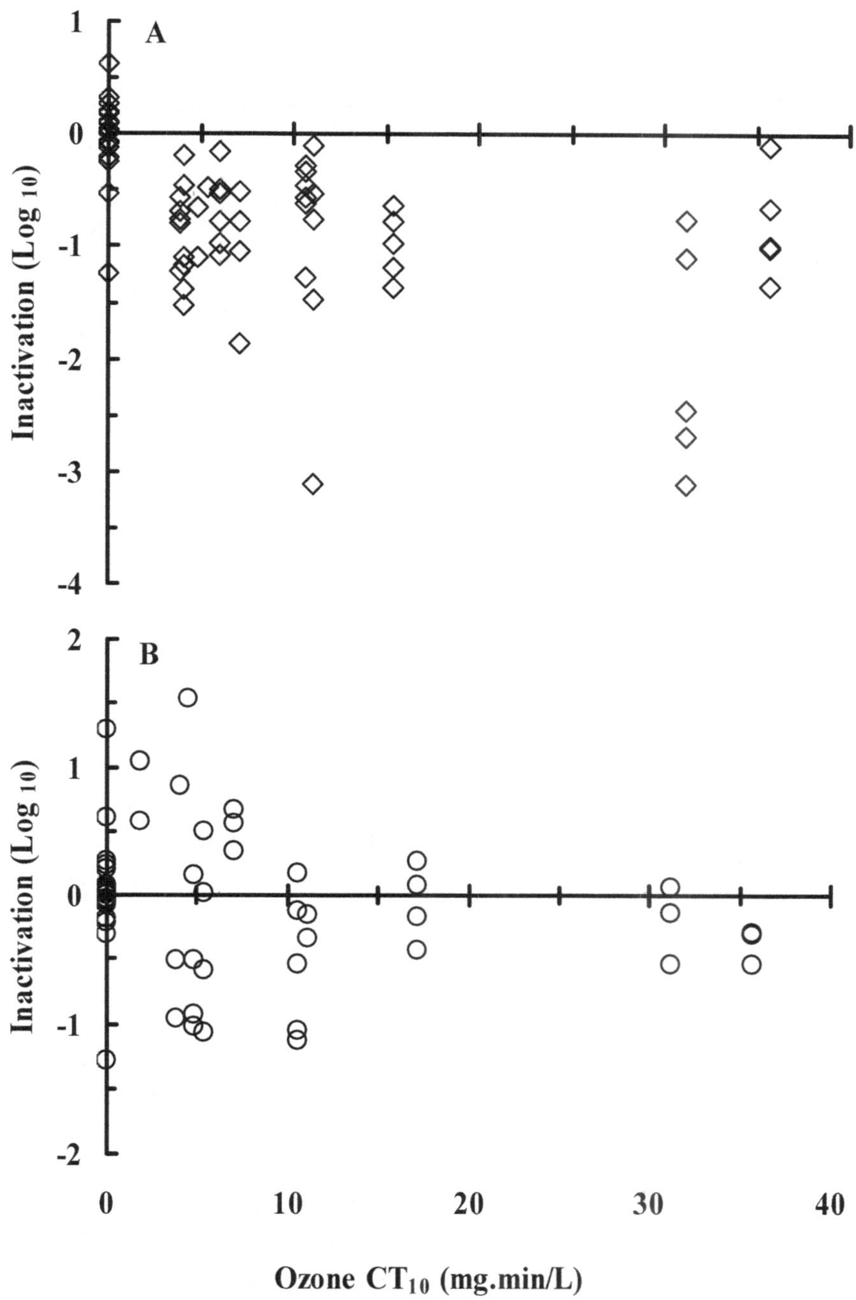

Figure 5.7 Inactivation of *C. parvum* in a continuously stirred tank reactor. Infectivity was measured in CD-1 mice (A) and HCT-8 cell culture (B) by solving the dose response model for the logit response obtained from each dose of ozone-exposed oocysts (approach 1 described above). Negative values on the y-axis indicate inactivation while positive values indicate stimulation.

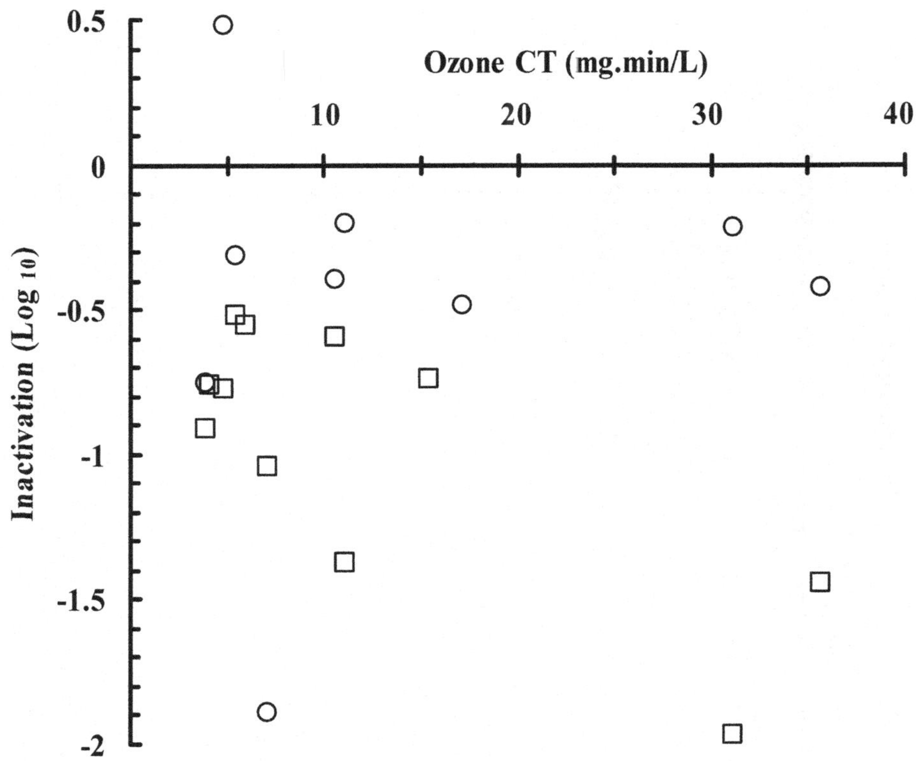

Figure 5.8 Inactivation of *C. parvum* in a continuously stirred tank reactor based on comparison of ID_{50} values. Infectivity was measured in CD-1 mice (□) and HCT-8 cell culture (○) by comparing the ID_{50}s of control and ozone-exposed oocysts (approach 2 described above).

Ozone Inactivation of *C. parvum* in Batch Reactors

Because the levels of inactivation in the CSTR experiments were lower than anticipated based on published literature, a series of experiments were performed using the simplest form of reactor, a batch system, which should generate the highest levels of inactivation. This was to confirm that that the general experimental set-up was appropriate, that ozone was being delivered to the target cells, and that the infectivity assays were performing as expected. The results demonstrated that 2-$\log_{10}$ inactivation levels were achieved at CT values of 13.7 and 12.4 mg.min/L when infectivity was measured by HCT-8 cell culture and CD-1 mice, respectively (Figure 5.9). For the limited set of data where direct comparisons were made, there was general agreement in the levels of inactivation measured by HCT-8 cell culture and CD-1 mice (Figure 5.10). These results, obtained at 15°C and 20°C, show good agreement with published data for ozone disinfection of *C. parvum* in batch reactors (Li et al., 2001; see Figure 1.5) and demonstrate that the overall experimental set-up was reliable. However, oocysts stimulation was also observed in these batch experiments at ozone doses less than 3 mg.min/L when infectivity was assessed by HCT-8 cell culture.

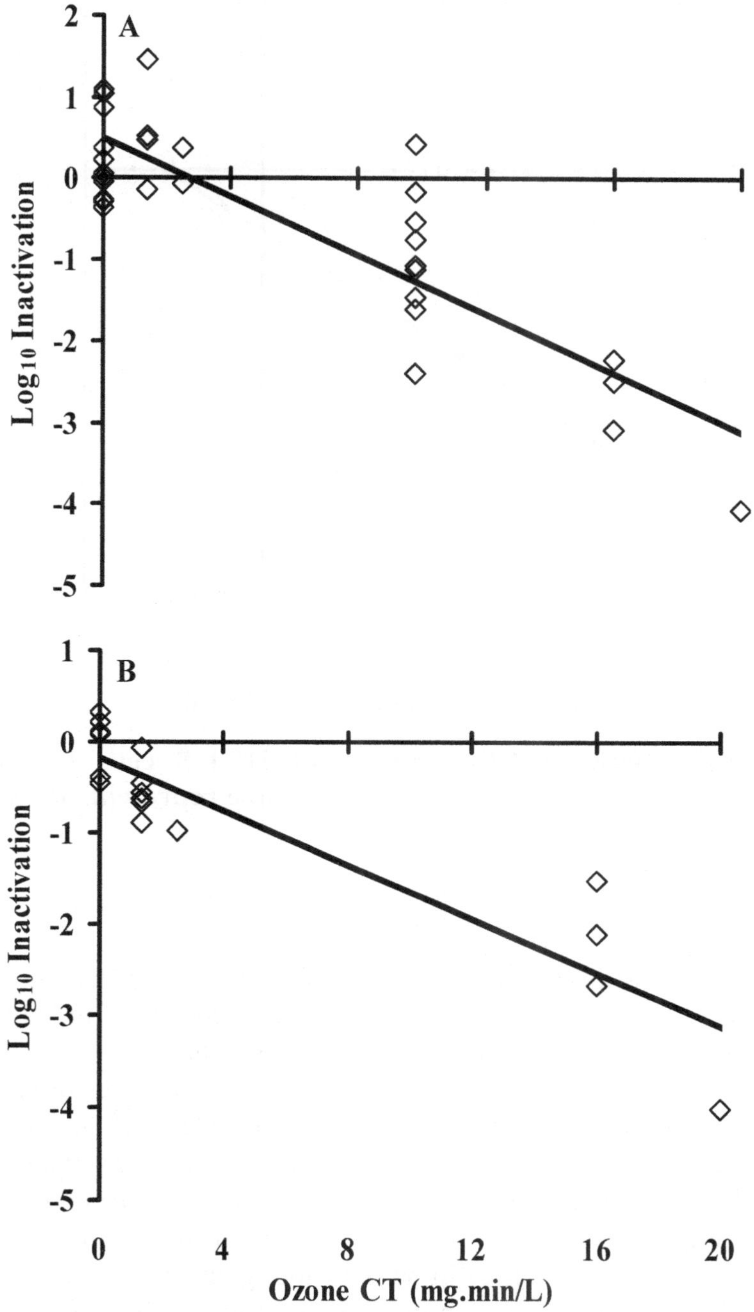

Figure 5.9 Ozone inactivation of *C. parvum* in a batch reactor. Infectivity was measured in HCT-8 cells (A) and CD-1 mice (B).

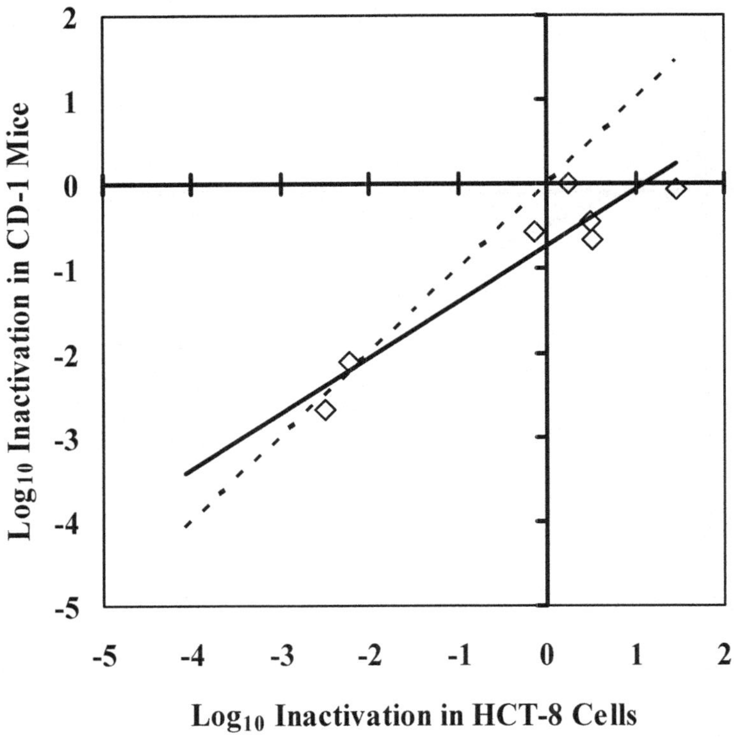

Figure 5.10 Correlation between CD-1 mice and HCT-8 cells for measuring ozone inactivation in a batch reactor (solid line). The dashed line represents the ideal correlation.

Semi-Batch Reactors

The majority of ozone disinfection experiments were performed in a 2 L semi-batch reactor as a compromise between the efficient inactivation achieved with batch reactors and the CSTR which is more representative of full-scale treatment plants.

A series of controls and tests were run to assess the physical performance of the reactor. Figure 5.11 displays the dissolved ozone and CT data from a semi-batch test using 0.05 M phosphate-buffered water (pH = 7, T = 15°C). Oocysts were added at t = 0 and the first dissolved ozone sample was taken shortly thereafter. While the ozone gas concentration and ozone gas flow rate were not varied, there was a ± ~0.1 mg/L variation in dissolved ozone concentration which was likely due to experimental error in both sampling and analysis. However, CT was not strongly influenced by the apparent variation in dissolved ozone. There was a linear relationship (R^2 = 0.99) between CT and time. During disinfection experiments, an a-priori CT estimate was used as the basis for collecting samples. The CT estimate was based on the steady-state dissolved ozone concentration measured before oocysts were added. A CT exposure time curve was then calculated, and samples were collected at times corresponding to the desired CT values. Once experiments were completed, an actual CT for each sample was then calculated. For most experiments, a-priori CT estimates closely approximated the measured CT values.

The Effect of the Oocyst Spike on the Ozone Demand and Decay Rate

An experiment was also conducted to determine whether the additions of *C. parvum* oocysts had an effect on the ozone demand and decay rate in the 2 L semi-batch reactor. Oocysts of the Iowa isolate (9 days post-shedding) were prepared according to the previously described procedures and the reactor was operated in a semi-batch mode where ozone gas was added and the dissolved ozone concentration was allowed to equilibrate before the beginning of each test. Two conditions, with and without oocysts, were tested in a 0.05-M phosphate buffered matrix (15°C, pH 7). Immediately after oocysts were added, ozone gas was shut off and grab samples were collected for dissolved ozone measurement. Figure 5.12 displays dissolved ozone concentrations for the two conditions tested. Differences in the initial data points show that the oocyst inocula did not increase ozone demand. The data follow a typical ozone decay curve, however, due to experimental error the profiles shown depart somewhat from an ideal logarithmic decay profile. Apparent first order ozone decay rates were calculated for each condition by plotting the natural logarithm of the normalized concentration against time (Figure 5.12B). Fitting a linear regression to the data allowed calculation of first order decay rates from the corresponding slope. First order decay rates were 0.05/min for the control condition (0.05 M phosphate buffer) and 0.054/min following addition of 1×10^8 oocysts to the reactor. Considering the precision of dissolved ozone data measurement, it can be concluded that the oocyst inocula did not measurably affect the dissolved ozone decay rate.

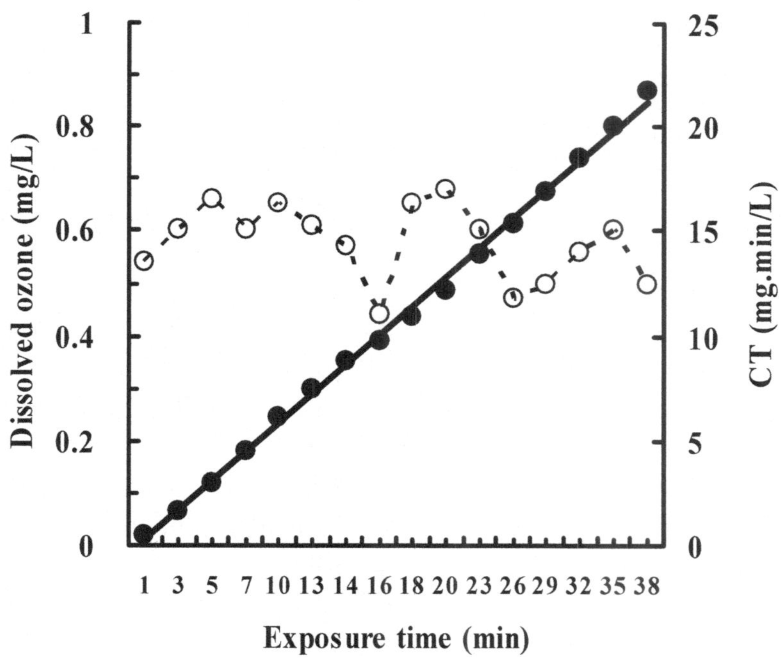

Figure 5.11 Dissolved ozone (○) and CT (•) measured in a typical 2 L semi-batch disinfection experiment.

Ozone Inactivation of *C. parvum* in Semi-Batch Reactors

Dose response data for control oocysts used in the semi-batch reactor are provided in Table 5.3. Unlike the earlier comparisons between cells and mice for measuring infectivity of untreated oocysts, there was generally poor agreement between the cell culture and mouse-derived ID_{50} values for control oocysts used in semi-batch experiments. Semi-batch experiments were performed at three temperatures (1°C, 5°C, and 15°C). Since temperature has a pronounced effect on the amount of inactivation achieved with ozone and because *C. parvum* oocysts may be damaged by freezing, the temperature within the reactor vessel was measured with a thermocouple linked to the chiller that circulated ethylene glycol through the jacket surrounding the reactor (Figure 5.13).

Simple linear regressions through the disinfection data generated using the CD-1 mouse model demonstrated that 3-$\log_{10}$ inactivation could be achieved with CT values of 14.9, 42.3, and 612.2 mg.min/L at 15°C, 5°, and 1°C, respectively (Figure 5.14). The CT values for 15°C and 5°C obtained with CD-1 mice in this study were in accordance with those published by the USEPA in the proposed LT2ESWTR, which requires ozone CTs of 19 and 47 mg.min/L to achieve 3-$\log_{10}$ inactivation at 15°C and 5°C, respectively (USEPA, 2003). This mouse-derived inactivation data demonstrated that oocysts were being exposed to ozone in the semi-batch reactor and that reasonable levels of inactivation could be achieved. Greater than 4-$\log_{10}$ inactivation was achieved at 15°C (Figure 5.14).

Table 5.3
Dose response data for control oocysts (CT = 0) used in semi-batch experiments

Trial	HCT-8 Cells Logit Dose Response	ID_{50}	CD-1 Mice Logit Dose Response	ID_{50}
1	LR = (1.790 × $\log_{10}$ dose) – 4.219	227	nd	
2	LR = (2.769 × $\log_{10}$ dose) – 5.391	88.5	LR = (0.732 × $\log_{10}$ dose) – 2.147	852
3	LR = (1.809 × $\log_{10}$ dose) – 4.406	273	nd	
4	LR = (4.032 × $\log_{10}$ dose) – 7.222	61.9	nd	
5	LR = (2.071 × $\log_{10}$ dose) – 5.197	323	LR = (3.002 × $\log_{10}$ dose) – 4.753	38.3
6	LR = (3.686 × $\log_{10}$ dose) – 6.021	43.0	Insufficient data*	
7	LR = (2.617 × $\log_{10}$ dose) – 5.043	84.6	LR = (3.670 × $\log_{10}$ dose) – 5.189	25.9
8	LR = (3.051 × $\log_{10}$ dose) – 4.354	26.7	nd	
9	LR = (2.374 × $\log_{10}$ dose) – 1.170	3.1	LR = (1.394 × $\log_{10}$ dose) – 2.201	37.9
10	LR = (8.728 × $\log_{10}$ dose) – 11.97	23.5	Insufficient data*	
11	LR = (2.145 × $\log_{10}$ dose) – 4.457	120	nd	
12	LR = (3.266 × $\log_{10}$ dose) – 3.483	11.7	LR = (1.977 × $\log_{10}$ dose) – 3.266	44.9
13	LR = (3.949 × $\log_{10}$ dose) – 3.743	8.9	nd	
14	LR = (3.818 × $\log_{10}$ dose) – 4.476	14.9	LR = (5.145 × $\log_{10}$ dose) – 9.368	66.2

* Multiple doses of oocysts were inoculated but most of them resulted in either 0% or 100% infection so the logit response could not be calculated.

nd, Not determined.

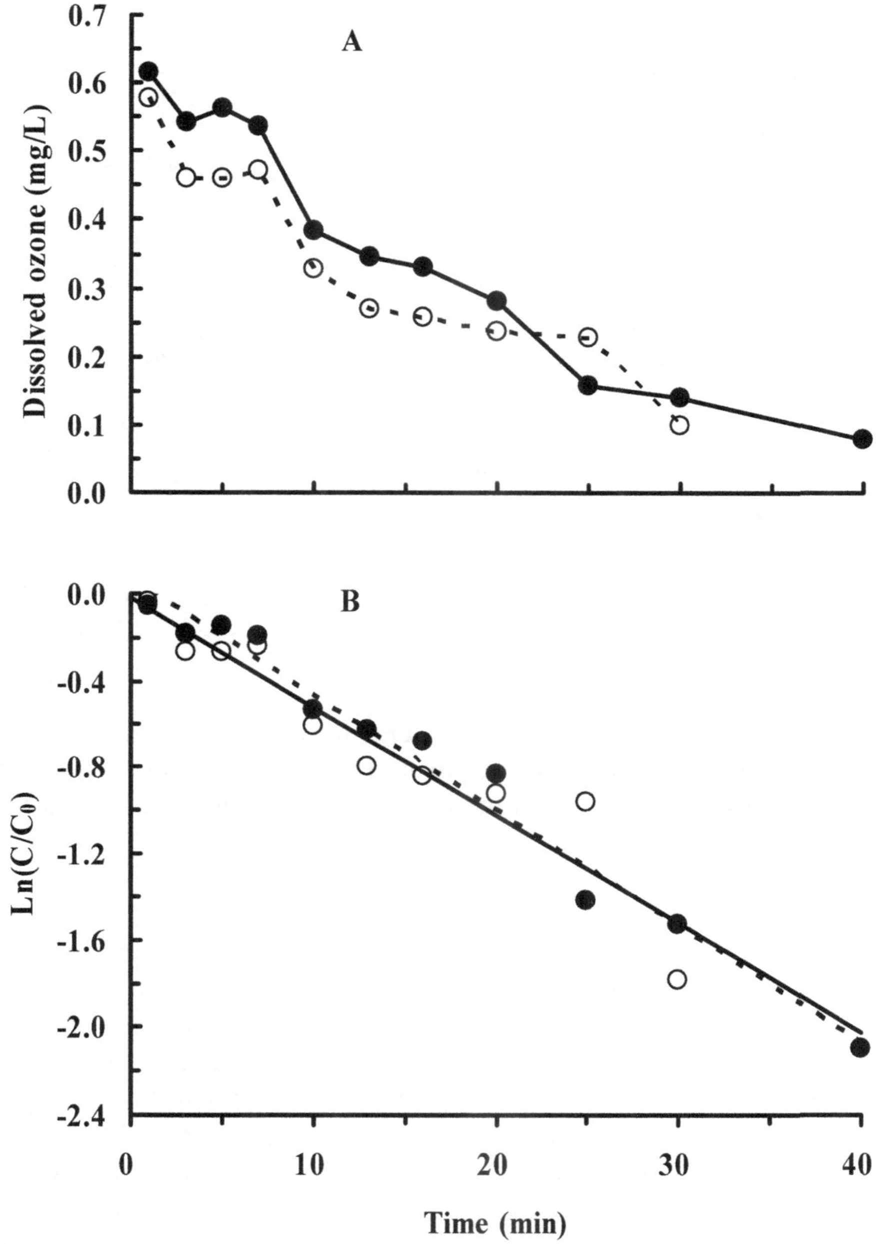

Figure 5.12 Effect of *C. parvum* oocysts on the ozone demand and decay in a 2 L semi-batch reactor plotted as dissolved ozone concentration (A) and as normalized ozone concentrations (B). Measurements were made at 15°C in 0.05 M phosphate buffer, pH 7.4 (○, dashed line, k = 0.050/min) and in phosphate buffer containing oocysts (•, solid line, k = 0.054/min).

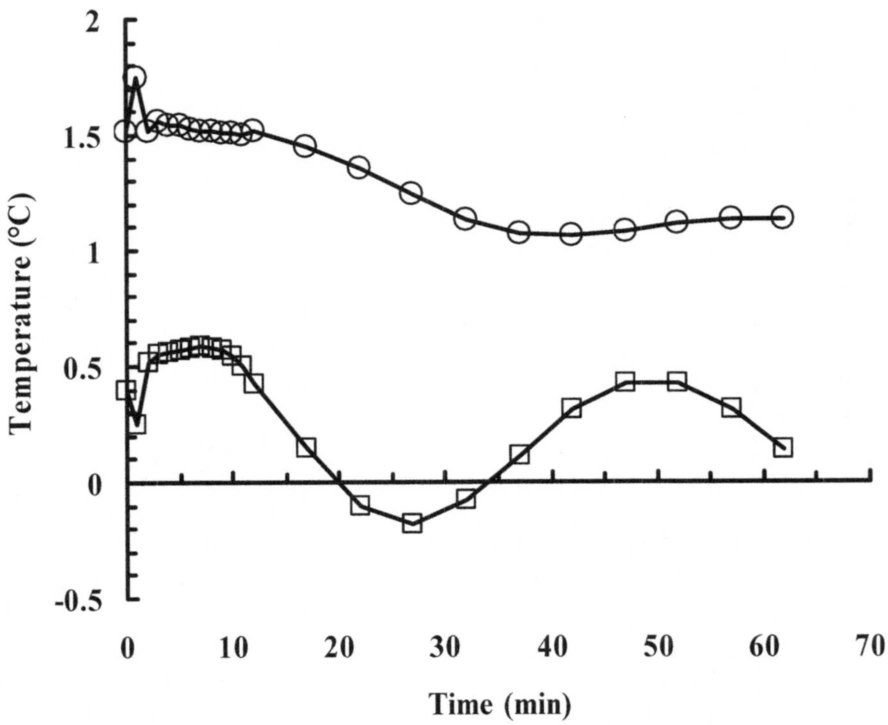

Figure 5.13 Temperature variations in the semi-batch ozone reactor during a 60 min run. The external temperature controller set to maintain the reactor temperature at 1°C. The internal reactor temperature (○) was measured with a thermocouple linked via a feedback loop to the temperature controller (□).

However, the inactivation levels obtained in this study were direct calculations from simple linear regressions through the data whereas the CT values included in the LT2ESWTR include a safety factor and are intended for application in full-scale treatment plants, which provide less efficient inactivation compared to bench-scale semi-batch reactors. Consequently, the data generated by the CD-1 mouse assay in this project should be considered as conservative relative to the raw data that was used to generate CT values in the LT2ESWTR.

The oocyst suspending matrices for the semi-batch reactors were 0.05 M phosphate buffer (pH 7) and untreated natural source water (pH range 8.18 – 8.4). Although only a limited amount of data was collected at pH 8 (all source waters were grouped together as pH 8), within the variability of the data, there was no difference between pH 7 and pH 8 data for inactivation at 5°C. The primary inactivation data was calculated by comparing actual oocyst dose and apparent oocyst dose based on solving the dose response equation for individual logit responses for each challenge dose of oocysts at each CT condition (Figure 5.14). Inactivation was also calculated by subtracting the ID_{50} value for ozone exposed oocysts from the control ID_{50}. However, this approach generated less data because it was not possible to construct dose response curves for all ozone-treated oocysts due to many 0% or 100% infectivity results (Figure 5.14). A wider range of oocyst dilutions would have circumvented this problem but the number of mice involved in each experiment would have been impractical.

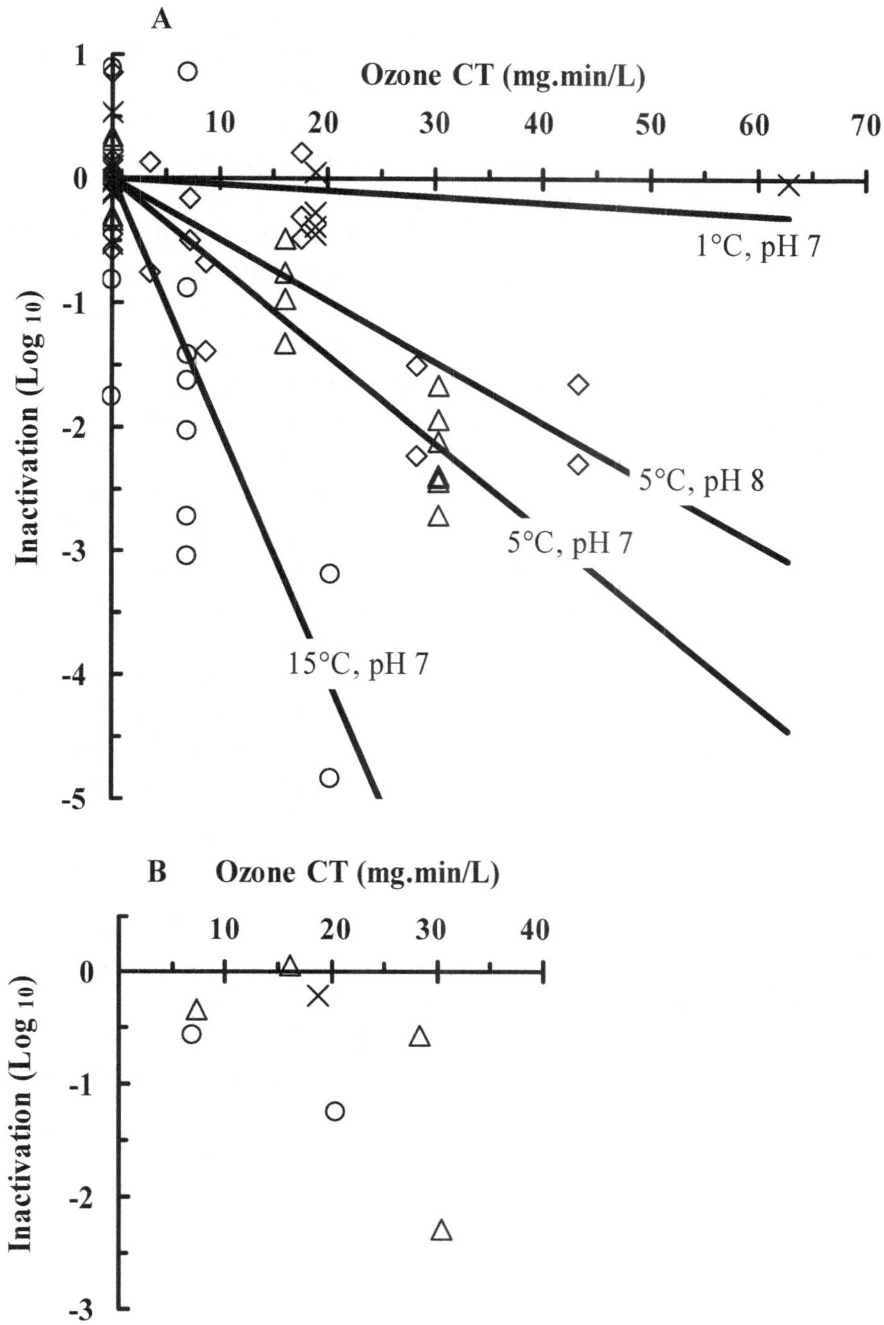

Figure 5.14 Inactivation of *C. parvum* by ozone in a semi-batch reactor assessed using the CD-1 mouse assay. Oocysts were exposed at 15°C (○), 5°C at pH 7 (Δ) and pH 8 (◊), and 1°C (×). Matrices were 0.05 M phosphate buffer at pH 7 and source water at pH 8. Inactivation was calculated by comparing actual oocyst dose and apparent oocyst dose based on solving the dose response equation for individual logit responses for each dose of oocysts (A) or by comparing ID_{50}s of treated and untreated oocysts (B).

Mouse derived inactivation data at 1°C were only obtained for ozone CTs of 18.8 and 62.8 mg.min/L. The single 62.8 mg.min/L result should probably be considered as an outlier due to experimental or assay variation since very little inactivation was observed, which was inconsistent with expectations based on the rest of the mouse semi-batch dataset. This was reflected in the extremely high calculated CT for 3-log_{10} inactivation of 612.2 mg.min/L. The mean inactivation at 18.8 mg.min/L was 0.41-log_{10}.

The HCT-8 cell culture-derived disinfection data was difficult to compare directly with the mouse data because some oocysts appeared to be stimulated by ozone when infectivity was assessed by cell culture (Figure 5.15). Although there was considerable scatter in the data, and some experiments resulted in oocyst inactivation at low CTs at both 15°C and 5°C, there were many datapoints that indicated that ozone exposed oocysts were more infectious than untreated, control oocysts (Figure 5.16). Based on a linear regression through the combined 15°C and 5°C data calculated using individual response logits (Figure 5.15A) and excluding control oocyst data, the transition from oocyst stimulation to oocyst inactivation occurred at a CT of approximately 20 mg.min/L. The nature of this stimulatory response is not yet understood but it appears to be unique to the cell culture model. However, investigating the mechanism and detailed kinetics of this unanticipated response was beyond the scope of the current project.

Increased infectivity (oocyst stimulation) was not observed when oocysts were exposed to ozone at 1°C (Figure 5.15). Based on linear regressions through the individual logit derived 1°C data, a CT of 85.5 mg.min/L was necessary to achieve 3-log_{10} inactivation (Figure 5.15A). When inactivation was calculated by comparing ID_{50} values for ozone-exposed and control oocysts, a CT of 66.2 mg.min/L was required to achieve 3-log_{10} inactivation at 1°C (Figure 5.15B). These values are in accordance with the CT in the proposed LT2ESWTR for 3-log_{10} inactivation at 1°C (69 mg.min/L) although, as explained above, the LT2ESWTR requirements include a safety factor.

The correlation between inactivation assessed using CD-1 mice and HCT-8 cell culture/RT-PCR for all ozone disinfection experiments where direct comparisons were made is displayed in Figure 5.17. Overall, the results demonstrated that the use of HCT-8 cell culture underestimated the level of inactivation relative to the CD-1 mouse assay. For the purposes of this comparison, inactivation was calculated as the mean inactivation for all oocyst challenge doses at each CT condition.

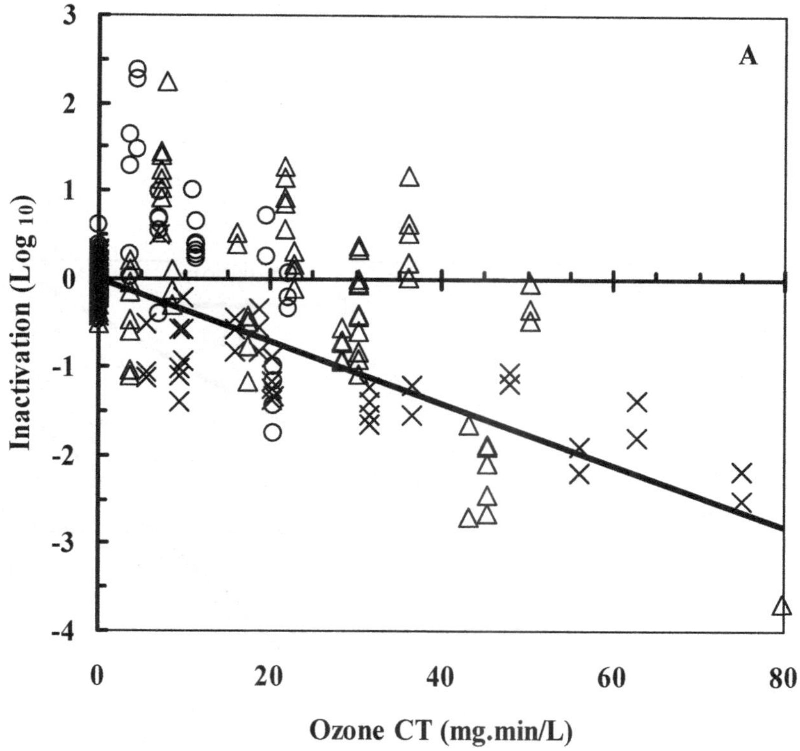

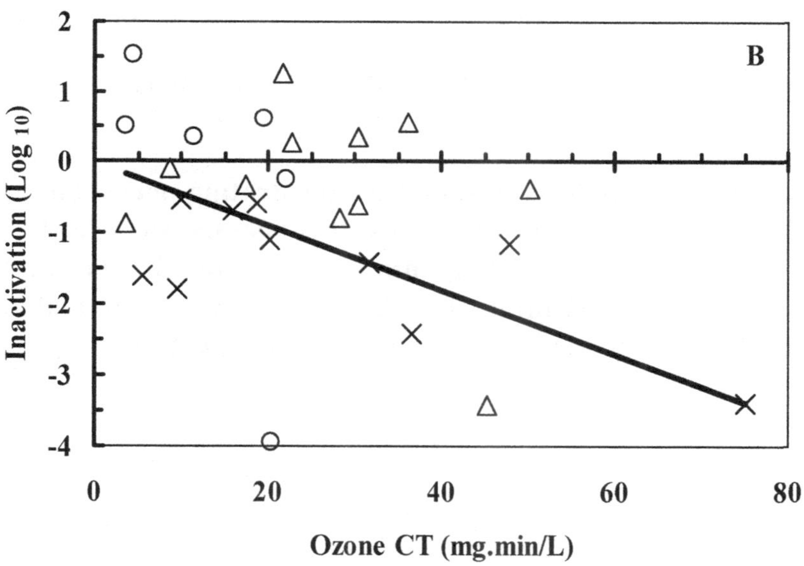

Figure 5.15 Inactivation of *C. parvum* by ozone in a semi-batch reactor assessed using the HCT-8/RT-PCR assay. Oocysts were exposed at 15°C (○), 5°C (Δ), and 1°C (×). Inactivation was calculated by comparing actual oocyst dose and apparent oocyst dose based on solving the dose response equation for individual logit responses for each dose of oocysts (A) or by comparing ID_{50}s of treated and untreated oocysts (B). A linear regression line was plotted through just the 1°C data.

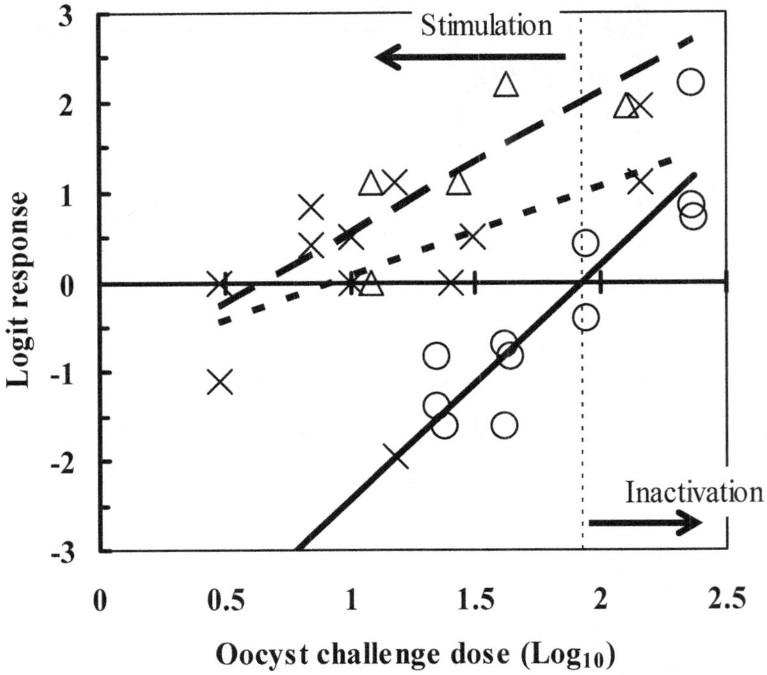

Figure 5.16 Example of a single experiment demonstrating oocyst stimulation in response to ozone exposure. The ID_{50} for untreated, control oocysts was 84.6 (○, —) compared to 4.5 for oocysts exposed to ozone at 7.3 mg.min/L (×, ----) and 8.7 for oocysts exposed to 21.8 mg.min/L of ozone (Δ, — — —), indicating that oocysts were stimulated by both of these conditions rather than inactivated. The experiment was performed at 5°C in source water (pH 8.4).

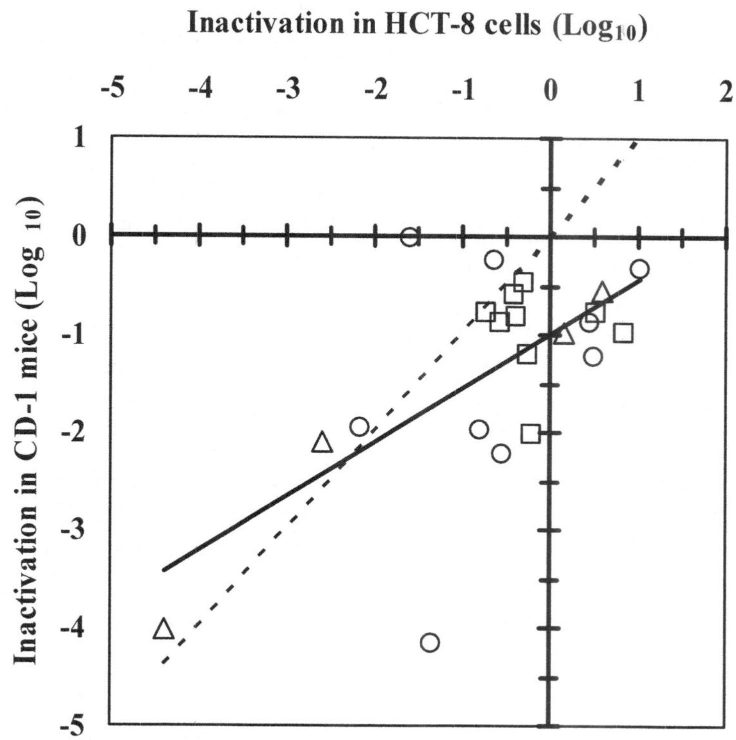

Figure 5.17 Comparison between HCT-8 cell culture/RT-PCR and CD-1 mouse assays for all ozone disinfection experiments that involved direct comparisons. Inactivation data were generated at 1°C, 5°C, and 15°C using batch (Δ), continuously stirred tank (□), and semi-batch (○) reactors. The dashed line represents the ideal correlation.

OOCYST STIMULATION IN RESPONSE TO OZONE EXPOSURE

Many of the ozone disinfection results generated using the HCT-8 cell culture/RT-PCR model appeared to be inconsistent with expectations based on previously published literature. Oocysts exposed to lower ozone CT values were more infectious than control oocysts. Since this seemed counter to expectations, an experiment was conducted specifically to address the potential for oocyst activation rather than inactivation at low ozone doses. Three challenge doses of oocysts suspended in 0.05 M phosphate buffer were exposed to two ozone CTs using a semi-batch reactor and inoculated onto HCT-8 monolayers. Dose response curves were constructed for untreated and ozone-exposed oocysts and inactivation was calculated by comparing the ID_{50} values for exposed and unexposed oocysts. Non-exposed control oocysts were added to 500 mL of 0.05 M phosphate buffer and held at 15°C for 30 min to match the conditions of exposed oocysts. Ozone-exposed and non-exposed oocysts were diluted in phosphate buffer and enumerated by IFA counts (5 counts per challenge dose). The ID_{50} for the untreated oocysts was 142 compared to 4 and 7 for oocysts exposed to ozone at CTs of 2.9 and 4.5, respectively. This corresponds to increases in infectivity of 1.6-$\log_{10}$ and 1.3-$\log_{10}$, respectively.

Infections were also quantitated using a colorimetric in-situ hybridization (CISH) assay. This assay was developed at MWD as a method to provide absolute quantitation of *C. parvum* infections in cell culture and utilizes a *C. parvum*-specific DNA probe to visualize infectious foci within a cell monolayer (Rochelle et al., 2001b). An infectious focus is defined as a cluster of *C. parvum* intracellular developmental stages (e.g. trophozoites, meronts, gamonts) that develop within a defined area of the cell monolayer resulting from cellular penetration by one or more sporozoites (Figure 5.18). Isolates or preparations of oocysts that are more infectious lead to the production of more infectious foci in cell culture compared to less infectious isolates or preparations. Based on the CISH assay, oocysts treated with low doses of ozone were also more infectious than untreated oocysts. Oocysts exposed to a CT of 2.9 mg.min/L produced an average of 15-fold more infectious foci compared to untreated oocysts while oocysts exposed to a CT of 4.5 mg.min/L produced an average of 6-fold more infectious foci compared to an equivalent number of untreated oocysts.

The use of two different infection detection methods demonstrated that the apparent stimulation was not an artifact of the RT-PCR detection methodology. The data from these two different methods of detecting and quantitating infection (RT-PCR and CISH) were combined to produce average oocyst stimulation levels of 1.3 ± 0.26-$\log_{10}$ at a CT of 2.9 mg.min/L and 0.97 ± 0.33-$\log_{10}$ at a CT of 4.5 mg.min/L. Combining data from this experiment with data from other semi-batch experiments reveals an overall trend of oocyst stimulation with ozone doses less than 8 mg.min/L.

While the phenomenon of oocyst stimulation at low ozone doses may initially seem counter-intuitive, it can be rationalized. Most of the early protocols for *C. parvum* infection in cell culture involved incubating oocysts in either bile salts (sodium taurocholate or sodium deoxycholate) and trypsin or 10% bleach (0.525% sodium hypochlorite) for 10 – 30 minutes prior to inoculation onto cell monolayers. Pretreatment of oocysts with sodium hypochlorite results in separation of the inner and outer oocyst walls enabling oocysts to excyst when warmed to 37°C, as happens when the oocysts are incubated following inoculation onto cell monolayers. Although untreated oocysts will also excyst when inoculated onto cell monolayers and incubated at 37°C, it is recognized that such oocyst pretreatments result in much higher levels of infection. Since bleach is an oxidant it may be expected that other oxidants, such as ozone, may have a stimulatory effect on oocyst infectivity in cell culture. Although the issue of oocyst stimulation by low doses of ozone has not been addressed directly in previous investigations of ozone disinfection, there are indications that other researchers have observed a similar phenomenon. Exposure to low doses of ozone increased the infectivity of *C. hominis* in MBDK cell monolayers compared to control oocysts and the authors hypothesized that ozone exposure may weaken the oocyst wall facilitating the release of sporozoites (Widmer et al., 2001). Also, exposure of *C. parvum* to hydrogen peroxide had a stimulatory effect on infectivity in bovine fallopian tube epithelial cells (Huang et al., 2001). In addition, it was reported that exposure of oocysts of the Iowa isolate to CT values in the range of 0.5 – 1.3 mg.min/L resulted in higher in-vitro excystation efficiencies compared to unexposed control oocysts, similar to the effect of pre-treating oocysts with sodium hypochlorite (Rennecker et al., 1999).

The reason that increased infectivity of ozone-exposed oocysts is not observed in mice is still to be resolved. One of the major differences between in-vitro cell culture and mouse infectivity assays is the conditions to which oocysts are exposed following inoculation into the infectivity model. Following exposure to ozone in this study, oocysts were shipped by overnight courier to the University of Arizona for inoculation into mice. Once inoculated, oocysts were immediately exposed to the harsh conditions of the mouse stomach (e.g., low pH, bile salts, various enzymes) before passing to the eventual site of infection in the intestines. Although oocysts may have been stimulated by low doses of ozone, the highly acidic conditions of the gut may be unfavorable to oocysts with separated inner and outer walls or to free sporozoites and therefore the oocysts do not have the opportunity to infect. In contrast, ozone-treated oocysts destined for cell culture were held overnight (to parallel shipment to UA) and then inoculated onto cell monolayers and incubated at 37°C. This immediately placed oocysts in an environment favorable to excystation and infection; ozone-stimulated oocysts had a chance to infect host cells without being exposed to acid or other conditions typical of the gut.

These results indicate that, while in-vitro cell culture is a sensitive and robust method for measuring the infectivity of untreated oocysts and for assessing the efficacy of some disinfectants, such as UV light, the particular assays used in this study, in their current form, may not be suitable for assessing the efficacy of oxidizing disinfectants. Consequently, there should be further research on the response of oocysts to oxidants when infection is measured by cell culture. Such experiments were beyond the scope of this project. Improvements or modifications to the method may include an oocyst treatment prior to cell inoculation to mimic the passage of oocysts through the stomach. Such modifications may result in cell culture becoming a more realistic model of in-vivo enteric environments, regardless of the handling or process conditions of the oocysts. Following the rationale that oxidant-treated oocysts are susceptible to inactivation by harsh stomach conditions, it is likely that the sequential application of a second oxidant following ozone exposure will result in oocyst inactivation, eliminating the stimulatory response. This theory should be tested in cell culture systems.

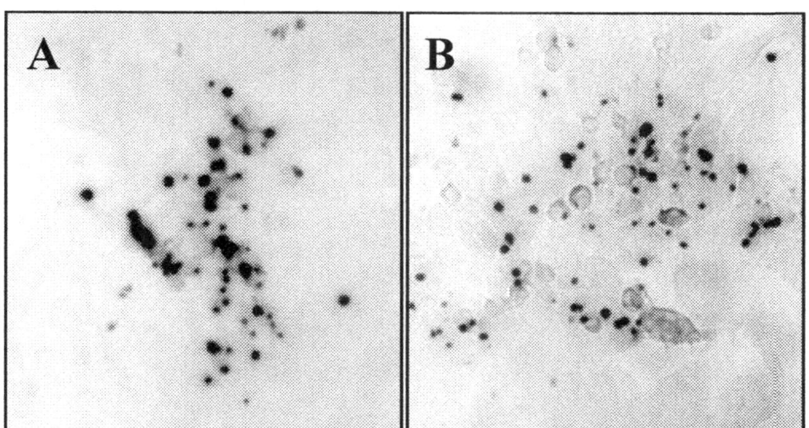

Figure 5.18 Detection of *C. parvum* infection in HCT-8 cells by colorimetric in-situ hybridization (CISH). Oocysts were either untreated (A) or exposed to ozone (2.9 mg.min/L) prior to inoculation of cell monolayers. Each panel displays individual infectious foci containing 50 – 68 parasite developmental stages per focus. The CISH probe was specific to *C. parvum* small subunit rRNA.

CHAPTER 6
SUMMARY AND CONCLUSIONS

The overall conclusion from this investigation is that in-vitro cell culture was equivalent to the "gold standard" of a mouse assay for measuring the infectivity of untreated *C. parvum* oocysts and should therefore be considered a practical alternative for assessing the potential of oocysts to cause infection. However, the high levels of variability displayed by mouse and cell culture methods indicated that infectivity and disinfection experiments should be limited to discerning relatively large differences. A cell culture-based assay was also demonstrated to be a practical method for assessing the efficacy of some disinfectants such as UV light. However, further research is necessary to understand the response of oocysts to ozone when inactivation is assessed by cell culture methods.

The specific conclusions of the project are:

1. There was good agreement between the CD-1 mouse assay and in-vitro cell culture for measuring the infectivity of fresh oocysts of five isolates of *C. parvum*. The general congruence was displayed in the shape of the dose response curves, the range of ID_{50} values, and the levels of variability.
2. Of the three cell culture assays used, infectivity results generated with the HCT-8/RT-PCR method were in closest agreement with results from the CD-1 mouse assay.
3. The ID_{50} values for the five *C. parvum* isolates in this study ranged from 13 to 629 for the various cell and mouse models. The mean and range of the ID_{50} values generated by the CD-1 mouse, HCT-8/RT-PCR cell culture, and Caco-2/RT-PCR cell culture assays, for all five isolates of *C. parvum* combined, were within the previously published 80% prediction limits of the Iowa isolate ID_{50}.
4. The wide variation in ID_{50} values for each isolate between the various assays makes it difficult to assign significance to the relative infectivity of different isolates of *C. parvum*. Nevertheless, the ID_{50} values obtained for the Iowa isolate were in general agreement with previous animal and human research and the TAMU and Moredun isolates appeared to be more infectious than other isolates.
5. The average standard deviation across all infectivity methods was 11.7%. Standard power analysis demonstrated that with three experimental replicates (10 mice or 10 wells of cell culture per replicate), the minimum difference in infectivity that could be detected was 23.7%. Reducing the detectable difference to 16% would require six replicates of 10 mice or 10 cell monolayers each (equivalent to a total of 60 mice or wells of cell culture) at each oocyst dose.
6. Infectivity differences between isolates may represent genuine differences in the levels of infectivity or they may simply be a reflection of combined sources of variability in the various assays.
7. Isolates of *C. hominis* were infectious in HCT-8 cell culture but did not infect mice.
8. Similar results were obtained with CD-1 mice and HCT-8 cell culture when measuring the infectivity of oocysts that had been exposed to an environmental water sample for 35 days.

9. There was very good agreement between the HCT-8 cell culture and CD-1 mouse assays for measuring UV inactivation of *C. parvum*. Combined data demonstrated that 2-$\log_{10}$ (99%) inactivation was achieved with a medium-pressure UV dosage of 5.6 mJ/cm^2.
10. The shapes of the ozone inactivation curves were generally the same for mouse and cell culture derived data although the CD-1 mouse assay typically generated 0.5 to 1-$\log_{10}$ higher levels of inactivation compared to the cell culture assay.
11. There was a pronounced stimulatory response in oocysts exposed to ozone below doses of 20 mg.min/L when assayed by HCT-8 cell culture. Sodium hypochlorite is a well known stimulant of oocysts and ozone appears to have similar properties when oocysts are inoculated into cell culture directly.
12. Up to 1.97-$\log_{10}$ inactivation was achieved in a single CSTR with a CT of 31.2 mg.min/L when infectivity was measured using CD-1 mice.
13. Using a batch reactor, at a CT of 16 mg.min/L, 2 – 3 $\log_{10}$ inactivation was achieved based on data derived from the CD-1 mouse and HCT-8 cell culture assays.
14. In a semi-batch reactor, 3-$\log_{10}$ inactivation was achieved at 14.9 and 42.3 mg.min/L at 15°C and 5°C, respectively, using CD-1 mice to measure infectivity.
15. Exposure to ozone at 1°C resulted in 3-$\log_{10}$ (99.9%) inactivation of the Iowa isolate of *C. parvum* with a CT of 66.2 mg.min/L as measured using the HCT-8 cell culture/RT-PCR assay.
16. The highest levels of ozone inactivation measured in this study were generally conservative relative to much of the published literature on ozone disinfection of *C. parvum* and comparison of all of the ozone disinfection data generated by this study indicated that overall the use of HCT-8 cell culture underestimated the level of inactivation relative to the CD-1 mouse assay.

RECOMMENDATIONS AND FUTURE RESEARCH

1. The water industry should adopt in-vitro cell culture as a routine method for measuring the infectivity of waterborne *C. parvum* and *C. hominis* oocysts. This project has demonstrated that cell culture has equivalency with the standard CD-1 mouse assay and cell culture assays can be applied to oocysts recovered from water using approved methods.
2. There needs to be a thorough, robust, and well-controlled study to compare the various cell culture-based assays for measuring *C. parvum* and *C. hominis* infectivity. This evaluation should include inter-laboratory comparisons and round-robin testing.
3. Cell culture-based assays should be used to assess the infectivity of a range of fresh *C. hominis* isolates. These isolates should either be recently recovered from infected humans or propagated using suitable animal models (e.g., gnotobiotic piglet).
4. Cell culture-based assays should be used for assessing the efficacy of selected disinfectants, such as ultraviolet light and other irradiation methods.
5. Modifications to cell culture growth conditions and infection detection methods should be evaluated to determine whether assay variability can be reduced. Efforts should also be made to automate all or parts of cell culture assays so that a large number of replicate analyses can be performed, which will maximize the confidence in the resultant data.
6. Further research should be conducted on the response of oocysts to oxidants such as ozone and chlorine when infection is measured in cell culture. The kinetics and mechanisms involved in the stimulatory response observed at the lower ozone CTs in this project need to be more fully understood.
7. In-vitro cell culture methods need to be adapted so that they represent a better model of in-vivo systems. These adaptations should involve processing oocysts to simulate passage through the stomach prior to inoculation of cell monolayers. Incubation in bile salts (sodium taurocholate) has been demonstrated to be beneficial but additional treatments may be necessary.
8. The use of cell culture to measure inactivation by the application of sequential disinfectants should be assessed. It is likely that the stimulatory response will be eliminated by sequential disinfection.
9. The response of *C. hominis* oocysts to disinfection should be assessed in cell culture.
10. There should be more research aimed at developing a cell culture-based method for propagating oocysts of *C. parvum* and *C. hominis*. Developments in the methodology that may increase the likelihood of success include low-shear rotating wall culture vessels, scaled-up culture volumes, serum-free culture media, and medium supplements. A successful method would eliminate the reliance on calves for propagating the large numbers of *C. parvum* oocysts needed for experimental purposes.

REFERENCES

Adams, R. B., R. L. Guerrant, S. X. Zu, G. D. Fang, and J. K. Roche. 1994. *Cryptosporidium parvum* infection of intestinal epithelium: morphologic and functional studies in an in vitro model. *Jour. Infect. Dis.*, 169:170-177.

Akiyoshi, D. E., X. Feng, M. A. Buckholt, G. Widmer, and S. Tzipori. 2002. Genetic analysis of a *Cryptosporidium parvum* human genotype 1 isolate passaged through different host species. *Infect. Immun.*, 70:5670-5675.

Akiyoshi, D. E., J. Dilo, C. Pearson, S. Chapman, J. Tumwine, and S. Tzipori. 2003. Characterization of *Cryptosporidium meleagridis* of human origin passaged through different host species. *Infect. Immun.*, 71:1828-1832.

Archer, J. R., J. R. Ball, J. H. Standridge, S. R. Greb, P. W. Rasmussen, J. P. Masterson, and L. Boushon. 1995. *Cryptosporidium* spp. oocyst and *Giardia* spp. cyst occurrence, concentrations and distribution in Wisconsin waters. Wisconsin Department of Natural Resources, Publication No. WR420-95.

Arrowood, M. J. 1988. Ph.D. thesis. University of Arizona, Tucson.

Arrowood, M. J., L. –T. Xie, and M. R. Hurd. 1994. In vitro assays of maduramicin activity against *Cryptosporidium parvum*. *Jour. Eukaryot. Microbiol.*, 41:23S.

Arrowood, M. J. and K. Donaldson. 1996. Improved purification methods for calf-derived *Cryptosporidium parvum* oocysts using discontinuous sucrose and cesium gradients. *Jour. Eukaryot. Microbiol.* 43:89S.

Arrowood, M. J. 2002. In vitro cultivation of *Cryptosporidium* species. *Clin. Microbiol. Rev.*, 15:390-400.

Belosevic, M., R. A. Guy, R. Taghi-Kilani, N. F. Neumann, L. L. Gyurek, L. R. J. Liyange, P. J. Millard, and G. R. Finch. 1997. Nucleic acid stains as indicators of *Cryptosporidium parvum* oocyst viability. *Int. Jour. Parasitol.*, 27:787-798.

Blewett, D. A., S. E. Wright, D. P., Casemore, N. E. Booth, and C. E. Jones. 1993. Infective dose size studies on *Cryptosporidium parvum* using gnotobiotic lambs. *Wat. Sci. Technol.*, 27:61-64.

Bonnin, A., I. Salimbeni, J. F. Dubremetz, G. Harly, P. Chavanet, and P. Camerlynck. 1990. Mise au point d'un modele experimental de culture in vitro des stades asexues de *Cryptosporidium* sp. *Ann. Parasitol. Hum. Comp.*, 65:41-43.

Brink, D. R., B. Langlais, and D. A. Reckhow. 1991. Introduction. In *Ozone in Water Treatment: Application and Engineering*, pp. 1-10. Eds. B. Langlais, D. A. Reckhow, and D. R. Brink. Lewis Publishers, Denver, CO.

Bukhari, Z., T. M. Hargy, J. R. Bolton, B. Dussert, and J. L. Clancy. 1999. Medium-pressure UV for oocyst inactivation. *Jour. AWWA*, 91(3):86-94.

Bukhari, Z., M. M. Marshall, D. G Korich, C. R. Fricker, H. V. Smith, J. Rosen, and J. L. Clancy. 2000. Comparison of *Cryptosporidium parvum* viability and infectivity assays following ozone treatment of oocysts. *Appl. Environ. Microbiol.*, 66:2972-2980.

Buraud, M., E. Forget, L. Favennec, J. Bizet, J. Gobert, and A. M. Deluol. 1991. Sexual stage development of cryptosporidia in the Caco-2 cell line. *Infect. Immunol.*, 59:4610-4613.

Caccio, S., W. Homan, R. Camilli, G. Traldi, T. Kortbeek, and E. Pozio. 2000. A microsatellite marker reveals population heterogeneity within human and animal genoptyes of *Cryptosporidium parvum*. *Parasitol.*, 120:237-244.

Campbell, A. T., L. J. Robertson, M. R. Snowball, and H. V. Smith. 1995. Inactivation of oocysts of *Cryptosporidium parvum* by ultraviolet irradiation. *Wat. Res.*, 29:2583-2586.

Carraway, M., S. Tzipori, and G. Widmer. 1997. A new restriction fragment length polymorphism from *Cryptosporidium parvum* identifies genetically heterogeneous parasite populations and genotypic changes following transmission from bovine to human hosts. *Infect. Immun.*, 65:3958-3960.

Carreno, R. A., D. S. Martin, and J. R. Barta. 1999. *Cryptosporidium* is more closely related to the gregarines than to coccidia as shown by phylogenetic analysis of apicomplexan parasites inferred using small subunit ribosomal RNA gene sequences from various hosts. *Parasitol. Res.*, 85:899-904.

Clesceri, L. S., A. E. Greenberg, and A. D. Eaton. 1998. *Standard Methods for the Examination of Water and Wastewater, 20th edition*. APHA, AWWA, WEF.

Communicable Disease Reports (CDR). 1999. Vol. 9 [Online]. Available: <http://www.phls.co.uk/publications/cdr.htm>.

Chappell, C. L., P. C. Okhuysen, C. R. Sterling, and H. L. DuPont. 1996. *Cryptosporidium parvum*: intensity of infection and oocyst excretion patterns in healthy volunteers. *Jour. Infect. Dis.*, 173:232-236.

Chauret, C. P., C. Z. Radziminski, M. Lepuil, R. Creason, and R. C. Andrews. 2001. Chlorine dioxide inactivation of *Cryptosporidium parvum* oocysts and bacterial spore indicators. *Appl. Environ. Microbiol.*, 67:2993-3001.

Chen, X. M., S. A. Levine, P. Tietz, E. Krueger, M. A. McNiven, D. M. Jefferson, M. Mahle, and N. F. LaRusso. 1998. *Cryptosporidium parvum* is cytopathic for cultured human biliary epithelia via an apoptotic mechanism. *Hepatology*, 28:906-913.

Clancy, J. L., T. M. Hargy, M. M. Marshall, and J. E. Dyksen. 1998. UV light inactivation of *Cryptosporidium* oocysts. *Jour. AWWA*, 90(9):92-102.

Clark, R. M., M. Sivagensen, E. W. Rice, and J. Chen. 2002. Development of a Ct equation for the inactivation of *Cryptosporidium* oocysts with ozone. *Wat. Res.*, 36:3141-3149.

Coffey, B. M., A. A. Mofidi, M. D. Williams, P. A. Rochelle, R. De Leon, and J. F. Green. 2001. Comparing UV and ozone disinfection of *Cryptosporidium parvum*: Implications for multi-barrier treatment. Proc. AWWA Annual Conference [available on CD-ROM], American Water Works Association, Denver, CO.

Craik, S. A., D. Weldon, G. R. Finch, J. Bolton, and M. Belosevic. 2001. Inactivation of *Cryptosporidium parvum* oocysts using medium- and low-pressure ultraviolet radiation. *Wat. Res.*, 35:1387-1398.

Current, W. L. and N. C. Reese. 1986. A comparison of endogenous development of three isolates of *Cryptosporidium* in suckling mice. *Jour. Protozool.*, 33:98-108.

Current, W. L. and T. B. Haynes. 1984. Complete development of *Cryptosporidium* in cell culture. *Science*, 224:603-605.

DeLaunay, A., G. Gargala, X. Li, L. Favennec, and J. J. Ballet. 2000. Quantitative flow cytometric detection of maximal *Cryptosporidium parvum* oocyst infectivity in a neonatal mouse model. *Appl. Environ. Microbiol.* 66:4315-4317.

Deng, M. Q., and D. O. Cliver. 1998. *Cryptosporidium parvum* development in the BS-C-1 cell line. *Jour. Parasitol.*, 84:8-15.

Di Giovanni, G. D., F. H. Hashemi, N. J. Shaw, F. A. Abrams, M. W. LeChevallier, and M. Abbaszadegan. 1999. Detection of infectious *Cryptosporidium parvum* oocysts in surface and filter backwash water samples by immunomagnetic separation and integrated cell culture PCR. *Appl. Environ. Microbiol.*, 65:3427-3432.

Driedger, A. M., J. L. Rennecker, and B. J. Marinas. 2001. Inactivation of *Cryptosporidium parvum* oocysts with ozone and monochloramine at low temperature. *Wat. Res.*, 35:41-48.

DuPont, H. L., C. L. Chappell, C. R. Sterling, P. C. Okhuysen, J. B. Rose, and W. Jakubowski. 1995. The infectivity of *Cryptosporidium parvum* in healthy volunteers. *N. Engl. Jour. Med.*, 332:885-859.

Facile, N., B. Barbeau, M. Prevost, and B. Koudjonou. 2000. Evaluating bacterial aerobic spores as a surrogate for *Giardia* and *Cryptosporidium* inactivation by ozone. *Wat. Res.*, 34:3238-3246.

Favennec, L., M. Ergaz-Bernard, E. Comby, D. Lemeteil, J. J. Ballet, and P. Brasseur. 1994. Immunofluorescence detection of *Cryptosporidium parvum* in Caco-2 cells: a new screening method for anticryptosporidial agents. *Jour. Eukaryot. Microbiol.*, 41:39S.

Fayer, R., C. A. Speer, and J. P. Dubey. 1990. General Biology of *Cryptosporidium*. In *Cryptosporidium of Man and Animals*, pp.2-29. Eds: J. P. Dubey, C. A. Speer, and R. Fayer. CRC Press, Boca Raton, Florida.

Fayer, R., C. A. Speer, and J. P. Dubey. 1997. The General Biology of *Cryptosporidium*. In *Cryptosporidium and Cryptosporidiosis*, pp.1-41. Ed: R. Fayer. CRC Press, Boca Raton, Florida.

Fayer, R., U. Morgan, and S. J. Upton. 2000. Epidemiology of *Cryptosporidium*: transmission, detection, and identification. *Int. Jour. Parasitol.*, 30:1305-1322.

Fayer, R., J. M. Trout, L. Xiao, U. M. Morgan, A. A. Lal, and J. P. Dubey. 2001. *Cryptosporidium canis* n. sp. from domestic dogs. *Jour. Parasitol.*, 87: 1415-1422.

Ferguson, D. M., R. De Leon, W. D. Taylor, and M. H. Stewart. 1998. Surface water monitoring in two watersheds to identify sources of *Cryptosporidium* spp. and *Giardia* spp. *Proc. Source Water Assessment and Protection 98*. Dallas. pp. 149-156.

Finch, G. R., C. W. Daniels, E. K. Black, F. W. Schaefer, III, and M. Belosevic. 1993. Dose-response of *Cryptosporidium parvum* in outbred, neonatal CD-1 mice. *Appl. Environ. Microbiol.*, 59:3661-3665.

Finch, G. R., L. R. J. Lyanage, and M. Belosevic. 1995. Effect of chlorine dioxide on *Cryptosporidium* and *Giardia*, p. 23-32. In *Symp. Proc. Chlorine Dioxide: Drinking Water, Process Water, and Wastewater Issues*. American Water Works Association, Denver, CO.

Finch, G. R., and K. Choe. 1999. Using *Bacillus* spores as a surrogate for ozone inactivation of *Cryptosporidium*. In *Proc. 14th Ozone World Congress International Ozone Association*, pp. 35-42. Dearborn, MI.

Flanigan, T. P., T. Aji, R. Marshall, R. Soave, M. Aikawa, and C. Kaetzel. 1991. Asexual development of *Cryptosporidium parvum* within a differentiated human enterocyte cell line. *Infect. Immun.*, 59:234-239.

Forney, J. R., S. Yang, C. Du, and M. C. Healey. 1996. Efficacy of serine protease inhibitors against *Cryptosporidium parvum* infection in a bovine fallopian tube epithelial cell culture system. *Jour. Parasitol.*, 82:638-640.

Gatei, W., R. W. Ashford, N. J. Beeching, S. K. Kamwati, J. Greensill, and C. A. Hart. 2002. *Cryptosporidium muris* infection in an HIV-infected adult, Kenya. *Emerg. Infect. Dis.*, 8:204-206.

Giacometti, A., O. Cirioni, M. Simona Del Prete, F. Barchiesi, A. Fineo, and G. Scalise. 2001. Activity of buforin II alone and in combination with azithromycin and minocycline against *Cryptosporidium parvum* in cell culture. *Jour. Antimicrob. Chemother.*, 47:97-99.

Griffiths, J. K. 1998. Human cryptosporidiosis: Epidemiology, transmission, clinical disease, treatment, and diagnosis. *Adv. Parasitol.*, 40:37-85.

Griffiths, J. K., R. Moore, S. Dooley, G. T. Keusch, and S. Tzipori. 1994. *Cryptosporidium parvum* infection of Caco-2 cell monolayers induces an apical monolayer defect, selectively increases transmonolayer permeability, and causes epithelial cell death. *Infect. Immun.*, 62:4506-4514.

Guerrant, D. I., S. R. Moore, A. A. Lima, P. D. Patrick, J. B. Schorling, and R. L Guerrant. 1999. Association of early childhood diarrhoea and cryptosporidiosis with impaired physical fitness and cognitive function. *Amer. Jour. Trop. Med. Hyg.*, 61:701-713.

Gut, J., C. Petersen, R. Nelson, and J. Leech. 1991. *Cryptosporidium parvum*: in vitro cultivation in Madin-Darby canine kidney cells. *Jour. Protozool.*, 38:72S-73S.

Gyurek, L. L., N. F. Neumann, G. R. Finch, and M. Belosevic. 2001. Comparison between animal infectivity and nucleic acid staining for determination of viability of ozone-inactivated *Cryptosporidium parvum* oocysts. *Ozone Sci. Engineer.*, 23:1-13.

Haas, C. N., and B. Kaymak. 2003. Effect of initial microbial density on inactivation of *Giardia muris* by ozone. *Wat. Res.*, 37:2980-2988.

Haas, C. N., K. French, G. R. Finch, and R. K. Guest. 2001. Data review on the physical/chemical removal of *Cryptosporidium*. Awwa Research Foundation and the American Water Works Association, Denver, CO.

Hayashi, K. 1994. Manipulation of DNA. In *The Polymerase Chain Reaction*, Eds: K. B. Mullis, F. Ferré, and R. Gibbs. Birkhäuser, Boston.

Hijjawi, N. S., B. P. Meloni, U. M. Morgan, and R. C. A. Thompson. 2001. Complete development and long-term maintenance of *Cryptosporidum parvum* human and cattle genotypes in cell culture. *Int. Jour. Parasitol.*, 31:1048-1055.

Hijjawi, N. S., B. P. Meloni, U. M. Ryan, M. E. Olson, and R. C. A. Thompson. 2002. Successful in vitro cultivation of *Cryptosporidum andersoni*: evidence for the existence of novel extracellular stages in the life cycle and implications for the classification of *Cryptosporidium*. *Int. Jour. Parasitol.*, 32:1719-1726.

Hirata, T., A. Shiumura, S. Morita, M. Suzuki, N. Motoyama, H. Hoshikawa, T. Moniwa, and M. Kaneko. 2001. The effect of temperature on the efficacy of ozonation for inactivating *Cryptosporidium parvum* oocysts. *Wat. Sci. Technol.*, 43:163-166.

Huang, K. H., S. G. Yang, and M. C. Healey. 2001. Effects of hydrogen peroxide on in vitro infection of *Cryptosporidium parvum*. *Prog. Nat. Sci.*, 11:946-951.

Hunter, P. R. and Q. Syed. 2001. Community surveys of self-reported diarrhea can dramatically overestimate the size of outbreaks of waterborne cryptosporidiosis. *Wat. Sci. Technol.*, 43:27-30.

Ionas, G., J. J. Learmouth, E. A. Keys, and T. J. Brown. 1998. Distribution of *Giardia* and *Cryptosporidium* in natural water systems in New Zealand- A nationwide survey. *Wat. Sci. Technol.*, 38:57-60.

Jenkins, M. B., L. J. Anguish, D. D. Bowman, M. J. Walker, and W. C. Ghiorse. 1997. Assessment of a dye permeability assay for determination of inactivation rates of *Cryptosporidium parvum* oocysts. *Appl. Environ. Microbiol.*, 63:3844-3850.

Khramtsov, N. V., P. A. Chung, C. C. Dykstra, J. K. Griffiths, U. M. Morgan, M. J. Arrowood, and S. J. Upton. 2000. Presence of double stranded RNAs in human and calf isolates of *Cryptosporidium parvum*. *Jour. Parasitol.*, 86:275-282.

Korich, D. G., J. R. Mead, M. S. Madore, N. A. Sinclair, and C. R. Sterling. 1990. Effects of ozone, chlorine dioxide, chlorine, and monochloramine on *Cryptosporidium parvum* oocyst viability. *Appl. Environ. Microbiol.*, 56:1423-1428.

Korich, D. G., M. M. Marshall, H. V. Smith, J. O'Grady, Z. Bukhari, C. R. Fricker, J. S. Rosen, and J. L. Clancy. 2000a. Inter-laboratory comparison of the CD-1 neonatal mouse logistic dose-response model for *Cryptosporidium parvum* oocysts. *Jour. Eukaryot. Microbiol.*, 47:294-298.

Korich, D. G., M. M. Marshall, and C. R. Sterling. 2000b. Lot-to-lot variation of *Cryptosporidium parvum* oocyst viability as assessed by the neonatal mouse infectivity assay. *In* Proceedings of the American Water Works Association Water Quality Technology Conference. American Water Works Association, Denver.

Larson, M. A., and B. J. Marinas. 2003. Inactivation of *Bacillus subtilis* spores with ozone and monochloramine. *Wat. Res.*, 37:833-844.

Lawton, P., M. Naciri, R. Mancassola, and A.-F. Petavy. 1996. Cultivation of *Cryptosporidium* in a non-adherent human monocytic cell line. *Jour. Microbiol. Meths.*, 27:165-173.

LeChevallier, M. W., and W. D. Norton. 1995. *Giardia* and *Cryptosporidium* in raw and finished water. *Jour. AWWA*, 87:54-68.

Li, H., G. R. Finch, D. W. Smith, and M. Belosevic. 2000. *Chemical Inactivation of Cryptosporidium in Water Treatment*. Awwa Research Foundation, Denver, CO.

Li, H., G. R. Finch, D. W. Smith, and M. Belosevic. 2001. Sequential inactivation of *Cryptosporidium parvum* using ozone and chlorine. *Wat. Res.*, 35:4339-4348.

Lindsay, D. S. 1997. Laboratory models of cryptosporidiosis, p.209-223. In R. Fayer (ed.), *Cryptosporidium and Cryptosporidiosis*. CRC Press, Boca Raton.

MacKenzie, W. R., N. J. Hoxie, M. E. Proctor, M. S. Gradus, K. A. Blair, D. E. Peterson, J. J. Kazmierczak, D. G. Addiss, K. R. Fox, J. B. Rose, and J. P. Davis. 1994. A massive outbreak in Milwaukee of *Cryptosporidium* infection transmitted through the public water supply. *New Engl. Jour. Med.*, 331:161-167.

Madore, M. S., J. B. Rose, C. P. Gerba, M. J. Arrowood, and C. R. Sterling. 1987. Occurrence of *Cryptosporidium* oocysts in sewage effluents and select surface waters. *Jour. Parasitol.*, 73:702-705.

Maillot, C., L. Favennec, A. Francois, P. Ducrotte, and P. Brasseur. 1997. Sexual and asexual development of *Cryptosporidium parvum* in five oocyst- or sporozoite-infected human enterocytic cell lines. *Jour. Eukaryot. Microbiol.*, 44:582-585.

Martinez, F., C. Mascaro, M. J. Rosales, J. Diaz, J. Cifuentes, and A. Osuna. 1992. In vitro multiplication of *Cryptosporidium parvum* in mouse peritoneal macrophages. *Vet. Parasitol.*, 42:27-31.

McLauchlin, J., C. Amar, S. Pedraza-Diaz, and G. L. Nichols. 2000. Molecular epidemiological analysis of *Cryptosporidium* spp. in the United Kingdom: Results of genotyping *Cryptosporidium* spp. in 1,705 fecal samples from humans and 105 fecal samples from livestock animals. *Jour. Clin. Microbiol.*, 38:3984-3990.

Messner, M. J., and R. L. Wolpert. 2003. *Cryptosporidium* and *Giardia* occurrence in ICR drinking water sources: Statistical analysis of ICR data. In *Information Collection Rule Data Analysis*, pp. 463-481. Eds. M. J. McGuire, J. L. McLain, and A. Obolensky. Awwa Research Foundation and the American Water Works Association, Denver, CO.

Millard, P. S., K. F. Gensheimer, and D. G. Addiss, 1994. An outbreak of cryptosporidiosis from fresh-pressed apple cider. *Jour. Amer. Med. Assoc.*, 272:1592-1596.

Mofidi, A. A., H. Baribeau, P. A. Rochelle, R. De Leon, B. M. Coffey, and J. F. Green. 2001. Disinfection of *Cryptosporidium parvum* with polychromatic UV light. *Jour. Amer. Wat. Works Assoc.*, 93(6):95-109.

Morgan, U., R. Weber, L. Xiao, I. Sulaiman, R. C. A. Thompson, W. Ndiritu, A. Lal, A. Moore, and P. Deplazes. 2000. Molecular characterization of *Cryptosporidium* isolates obtained from human immunodeficiency virus-infected individuals living in Switzerland, Kenya, and the United States. *Jour. Clin. Microbiol.*, 38:1180-1183.

Morgan-Ryan, U., A. Fall, L. A. Ward, N. Hijjawi, I. Sulaiman, R. Fayer, R. C. Thompson, M. Olson, A. Lal, and L. Xiao. 2002. *Cryptosporidium hominis* n. sp. (Apicomplexa: Cryptosporidiidae) from *Homo sapiens*. *Jour. Eukaryot. Microbiol.*, 49:433-440.

Naciri, M., P. Yvore, C. de Boissieu, and E. Esnault. 1986. Multiplication de *Cryptosporidium muris* (Tyzzer 1907) in vitro entretien d'une souche sur oeufs embryonnes. *Rec. Med. Vet.*, 162:51-56.

National Academy of Sciences. 1996. *Guide for the care and use of laboratory animals*. National Academy Press, Washington, D. C.

Nickerson, C. A., T. J. Goodwin, J. Terlonge, C. M. Ott, K. L. Buchanan, W. C. Uicker, K. Emami, C. L. LeBlanc, R. Ramamurthy, M. S. Clarke, C. R. Vanderburg, T. Hammond, and D. L. Pierson. 2001. Three-dimensional tissue assemblies: novel models for the study of *Salmonella enterica* serovar Typhimurium pathogenesis infection. *Infect. Immun.*, 69:7106-7120.

Okhuysen, P. C., C. L. Chappell, C. R. Sterling, W. Jakubowski, and H. I DuPont. 1998a. Susceptibility and serologic response of healthy adults to reinfection with *Cryptosporidium parvum*. *Infect. Immun.*, 66:441-443.

Okhuysen P. C., C. L. Chappell, J. Crabb, L. M. Valdez, E. T. Douglas, and H. I DuPont. 1998b. Prophylactic effect of bovine anti-*Cryptosporidium* hyperimmune colostrum immunoglobulin in healthy volunteers challenged with *Cryptosporidium parvum*. *Clin. Infect. Dis.*, 26:1324-1329.

Okhuysen, P. C., C. L. Chappell, J. H. Crabb, C. R. Sterling, and H. L. DuPont. 1999. Virulence of three distinct *Cryptosporidium parvum* isolates for healthy adults. *Jour. Infect. Dis.*, 180:1275-1281.

Okhuysen, P. C., S. M. Rich, C. L. Chappell, K. A. Grimes, G. Widmer, X. Feng, and S. Tzipori. 2002. Infectivity of a *Cryptosporidium parvum* isolate of cervine origin for healthy adults and interferon-γ knockout mice. *Jour. Infect. Dis.*, 185:1320-1325.

Oppenheimer, J. A., E. M. Aieta, R. R. Trussell, J. G. Jacangelo, and I. Najm. 2000. Evaluation of *Cryptosporidium* inactivation in natural waters. AWWA Research Foundation and the American Water Works Association, Denver, CO.

Owens, J. H., R. J. Miltner, T. R. Slifko, and J. B. Rose. 1999. In vitro excystation and infectivity in mice and cell culture to assess chlorine dioxide inactivation of *Cryptosporidium parvum*. In *Proc. AWWA Water Quality Technology Conference*, Denver, CO.

Owens, J. H., R. J. Miltner, E. W. Rice, C. H. Johnson, D. R. Dahling, F. W. Schaeffer, and H. M. Shukairy. 2000. Pilot-scale ozone inactivation of *Cryptosporidium* and other microorganisms in natural water. *Ozone Sci. Engineer.*, 22:501-517.

Pedraza-Diaz, S., C. Amar, A. M. Iversen, P. J. Stanley, and J. McLauchlin. 2001. Unusual *Cryptosporidium* species recovered from human faeces: first description of *Cryptosporidium felis* and *Cryptosporidium* 'dog type' from patients in England. *Jour. Med. Microbiol.*, 50:293-296.

Peng, M. M., L. Xiao, A. R. Freeman, M. J. Arrowood, A. A. Escalante, A. C. Weltman, C. S. L Ong, W. R. Mackenzie, A. A. Lal, and C. B. Beard. 1997. Genetic polymorphism among *Cryptosporidium parvum* isolates: Evidence of two distinct human transmission cycles. *Emerg. Infect. Dis.*, 3:567-573.

Pieniazek, N. J., F. J. Bornay-Llinares, S. B. Slemenda, A. J. da Silva, I. N. Moura, M. J. Arrowood, O. Ditrich, and D. G. Addiss. 1999. New *Cryptosporidium* genotypes in HIV-infected persons. *Emerg. Infect. Dis.*, 5:444-449.

Pinto, M., L. S. Robine, M. D. Appay, M. Kedinger, N. Traidou, E. Dussault, B. Lacroix, A. P. Simon, K. Haffen, J. Fogh, and A. Zweibaum. 1983. Enterocyte like differentiation and polarization of the human colon carcinoma cell line Caco-2 in cell culture. *Biol. Cell.* 47:323-330.

Rasmussen, K. R., N. C. Larsen, and M. C. Healey. 1993. Complete development of *Cryptosporidium parvum* in a human endometrial carcinoma cell line. *Infect. Immun.*, 61:1482-1485.

Rennecker, J. L., B. J. Marinas, J. H. Owens, and E. W. Rice. 1999. Inactivation of *Cryptosporidium parvum* oocysts with ozone. *Wat. Res.*, 33:2481-2488.

Rennecker, J. L., A. M. Driedger, S. A. Rubin, and B. J. Marinas. 2000. Synergy in sequential inactivation of *Cryptosporidium parvum* with ozone/free chlorine and ozone/monochloramine. *Wat. Res.*, 34:4121-4130.

Rennecker, J. L, J. H. Kim, B. Corona-Vasquez, and B. J. Marinas. 2001. Role of disinfectant concentration and pH in the inactivation kinetics of *Cryptosporidium parvum* oocysts with ozone and monochloramine. *Environ. Sci. Technol.*, 35:2752-2757.

Rochelle, P. A. 2001. Detection of protozoa in environmental water samples. In *Environmental Molecular Microbiology: Protocols and Applications*, pp. 91- 114. Ed. P. A. Rochelle. Horizon Scientific Press, Wymondham.

Rochelle, P. A., and R. De Leon. 2001. A review of methods for assessing the infectivity of *Cryptosporidium parvum* using in-vitro cell culture. In *Cryptosporidium: The Analytical Challenge*, pp. 88-95. Eds. M. Smith and K. C. Thompson. Royal Society of Chemistry, Cambridge.

Rochelle, P. A., D. M. Ferguson, T. J. Handojo, R. De Leon, M. H. Stewart, and R. L. Wolfe. 1997. An assay combining cell culture with reverse transcriptase-PCR to detect and determine infectivity of waterborne *Cryptosporidium parvum*. *Appl. Environ. Microbiol.*, 63:2029-2037.

Rochelle, P. A., R. De Leon, A. Johnson, M. H. Stewart, and R. L. Wolfe. 1999a. Evaluation of immunomagnetic separation for recovery of infectious *Cryptosporidium parvum* oocysts from environmental samples. *Appl. Environ. Microbiol.*, 65:841-845.

Rochelle, P. A., E. M. Jutras, E. R. Atwill, R. De Leon, and M. H. Stewart. 1999b. Polymorphisms in the β-tubulin gene of *Cryptosporidium* parvum differentiate between isolates based on animal host but not geographic origin. *Jour. Parasitol.*, 85:986-989.

Rochelle, P. A., R. De Leon, and E. R. Atwill. 2001a. *Evaluation of Genotyping Techniques for Cryptosporidium parvum*. Awwa Research Foundation and the American Water Works Association, Denver, CO.

Rochelle, P. A., D. M. Ferguson, A. M. Johnson, and R. De Leon. 2001b. Quantitation of *Cryptosporidium parvum* infection in cell culture using a colorimetric in situ hybridization assay. *Jour. Eukaryot. Microbiol.* 48:565-574.

Rochelle, P. A., Marshall, M. M., Mead, J. R., Johnson, A. M., Korich, D. G., Rosen, J. S., and De Leon, R. 2002. Comparison of in vitro cell culture and a mouse assay for measuring infectivity of *Cryptosporidium parvum*. *Appl. Environ. Microbiol.*, 68:3809-3817.

Rochelle, P. A., A. A. Mofidi, M. M. Marshall, S. J. Upton, B. Montelone, and K. Woods. 2004. *An Investigation of UV Disinfection and Repair in Cryptosporidium parvum*. Awwa Research Foundation, Denver, CO.

Rose, J. B., J. T. Lisle, and M. LeChevallier. 1997. Waterborne cryptosporidiosis: Incidence, outbreaks, and treatment strategies, p.93-109. *In* R. Fayer (ed.), *Cryptosporidium* and Cryptosporidiosis. CRC Press, Boca Raton.

Rosen, J. S., P. A. Rochelle, M. Marshall, J. Sobrinho, D. Korich, J. R. Mead, and R. De Leon. 2001. Comparative study of variability in mouse infectivity and cell culture/RT-PCR assays. *Proc. AWWA Water Quality Technology Conference*. [on CD-ROM]. Denver, Colo.: AWWA.

Ruffell, K. M., J. L. Rennecker, and B. J. Mariñas. 2000. Inactivation of *Cryptosporidium parvum* oocysts with chlorine dioxide. *Wat. Res.*, 34:868-876.

Sambrook, J., P. MacCallum, and D. Russell. 2001. *Molecular Cloning: A Laboratory Manual*. Cold Spring Harbor, N.Y.: Cold Spring Harbor Laboratory Press.

Shin, G.-A., K. G., Linden, M. J., Arrowood, and M. D. Sobsey, 2001. Low-pressure UV inactivation and DNA repair potential of *Cryptosporidium parvum*. *Appl. Environ. Microbiol.*, 67:3029-3032.

Slifko, T. R., D. Friedman, J. B. Rose, and W. Jakubowski. 1997. An in vitro method for detecting infectious *Cryptosporidium* oocysts with cell culture. *Appl. Environ. Microbiol.*, 63:3669-3675.

Slifko, T. R., D. E. Huffman, and J. B. Rose. 1999. A most-probable-number assay for enumeration of infectious *Cryptosporidium parvum* oocysts. *Appl. Environ. Microbiol.*, 65:3936-3941.

Spano, F., L. Putignani, J. McLauchlin, D. P. Casemore, and A. Crisanti. 1997. PCR-RFLP analysis of the *Cryptosporidium* oocyst wall protein (COWP) discriminates between *C. wrairi* and *C. parvum*, and between *C. parvum* isolates of human and animal origin. *FEMS Microbiol. Lett.*, 150:209-217.

Spano, F., L. Putignani, A. Crisanti, P. Sallicandro, U. Morgan, S. M. Le Blancq, L. Tchack, S. Tzipori, and G. Widmer. 1998. Multilocus genotypic analysis of *Cryptosporidium parvum* isolates from different hosts and geographical origins. *Jour. Clin. Microbiol.*, 36:3255-3259.

States, S., K. Stadterman, L. Ammon, P. Vogel, J. Baldizar, D. Wright, L. Conley, and J. Sykora. 1997. Protozoa in river water: sources, occurrence, and treatment. *Jour. AWWA*, 89:74-83.

Stinear, T., A. Matusan, K. Hines, and M. Sandery. 1996. Detection of a single viable *Cryptosporidium parvum* oocyst in environmental water concentrates by reverse trasnscription-PCR. *Appl. Environ. Microbiol.*, 62:3385-3390.

Sulaiman, I. M., L. Xiao, C. Yang, L. Escalante, A. Moore, C. B. Beard, M. J. Arrowood, and A. A. Lal. 1998. Differentiating human from animal isolates of *Cryptosporidium parvum*. *Emerg. Infect. Dis.*, 4:681-685.

U. S. Environmental Protection Agency (USEPA). 1999. Method 1623: *Cryptosporidium* and *Giardia* in Water by Filtration/IMS/FA. EPA 821-R-99-006.

U. S. Environmental Protection Agency (USEPA). 2003. National Primary Drinking Water Regulations: Long Term 2 Enhanced Surface Water Treatment Rule; Proposed Rule. Federal Register, 40 CFR Parts 141 and 142, Vol. 68, No. 154

Upton, S. J., M. Tilley, M. V. Nesterenko, and D. B. Brillhart. 1994. A simple and reliable method of producing in vitro infections of *Cryptosporidium parvum* (Apicomplexa). *FEMS Microbiol. Letts.*, 118:45-50.

Upton, S. J. 1997. In vitro cultivation. In *Cryptosporidium and Cryptosporidosis*, pp. 181-207. Ed: R. Fayer. CRC Press, Boca Raton, Florida.

Vesey, G., N. Ashbolt, E. J. Fricker, D. Deere, K. L. Williams, D. A. Veal, and M. Dorsch. 1998. The use of a ribosomal RNA targeted oligonucleotide probe for fluorescent labeling of viable *Cryptosporidium parvum* oocysts. *Jour. Appl. Microbiol.*, 85:429-440.

Wallis, P. M., S. L. Erlandsen, J. L. Isaac-Renton, M. E. Olson, W. J. Robertson, and H. van Keulen. 1996. Prevalence of *Giardia* cysts and *Cryptosporidium* oocysts and characterization of *Giardia* spp. cysts isolated from drinking water in Canada. *Appl. Environ. Microbiol.*, 62:2789-2797.

White, G. C. 1999. *Handbook of Chlorination and Alternative Disinfectants*. John Wiley and Sons, Ltd.

Widmer, G., E. A. Orbacz, and S. Tzipori. 1999. β-tubulin mRNA as a marker of *Cryptosporidium parvum* oocyst viability. *Appl. Environ. Microbiol.*, 65:1584-1588.

Widmer, G., D. Akiyoshi, M. A. Buckholt, X. Feng, S. M. Rich, K. M. Deary, C. A. Bowman, P. Xu, Y. Wang, X. Wang, G. A. Buck, and S. Tzipori. 2000a. Animal propagation and genomic survey of a genotype 1 isolate of *Cryptosporidium parvum*. *Mol. Biochem. Parasitol.* 108:187-197.

Widmer, G., E. A. Corey, B. Stein, J. K. Griffiths, and S. Tzipori. 2000b. Host cell apoptosis impairs *Cryptosporidium parvum* development in vitro. *Jour. Parasitol.*, 86:922-928.

Widmer, G., H. Ward, Z. Bukhari, T. Clancy, J. Crabb, and S. Tzipori. 2001. Molecular mechanisms of *Cryptosporidium* oocyst and *Giardia* cyst inactivation by chemical disinfectants. *Proc. AWWA Water Quality Technology Conference*. [on CD-ROM]. Denver, Colo.: AWWA.

Woods, K. M., M. V. Nesterenko, and S. J. Upton. 1995. Development of a microtiter ELISA to quantify development of *Cryptosporidium parvum* in vitro. *FEMS Microbiol. Letts.*, 128:89-94.

Xiao, L., L. Escalante, C. Yang, I. Sulaiman, A. A. Escalante, R. J., Montali, R. Fayer, and A. A. Lal. 1999. Phylogenetic analysis of *Cryptosporidium* parasites based on the small-subunit rRNA gene locus. *Appl. Environ. Microbiol.*, 65:1578-1583.

Xiao, L., K. Alderisio, J. Limor, M. Royer, and A. A. Lal. 2000. Identification of species and sources of *Cryptosporidium* oocysts in storm waters with a small-subunit rRNA-based diagnostic and genotyping tool. *Appl. Environ. Microbiol.*, 66:5492-5498.

Xiao, L., C. Bern, J. Limor, I. Sulaiman, J. Roberts, W. Checkley, I. Cabrera, R.H. Gilman, and A.A. Lal. 2001. Identification of 5 types of *Cryptosporidium* parasites in children from Lima, Peru. *Jour. Infect. Dis.*, 185:270-271.

Yang, S., M. C. Healey, C. Du, and J. Zhang. 1996. Complete development of *Cryptosporidium parvum* in bovine fallopian tube epithelial cells. *Infect. Immun.*, 64:349-354.

You, X., M. J. Arrowood, M. Lejkowski, L. Xie, R. F. Schinazi, and J. R. Mead. 1996. A chemiluminescence immunoassay for evaluation of *Cryptosporidium parvum* growth in vitro. *FEMS Microbiol. Letts.*, 136:251-256.

ABBREVIATIONS

Abs	Antibiotics
ANCOVA	Analysis of covariance
ATCC	American Type Culture Collection
AWWA	American Water Works Association
AwwaRF	Awwa Research Foundation
bp	Base pairs
BSA	Bovine serum albumin
°C	Degrees Celsius
CDR	Communicable Disease Reports
CI	Confidence interval
CISH	Colorimetric in-situ hybridization
cm	Centimeter
cm^2	Square centimeters
CO_2	Carbon dioxide
COWP	*Cryptosporidium* oocyst wall protein
CRW	Colorado River Water (source water)
CSTR	Continuously stirred tank reactor
CT	Concentration (mg/L) × time (minutes) product
CV	Coefficient of variation
d	Days
DAPI	4',6'-diamidino-2-phenylindole
dATP	2'-deoxyadenosine 5'-triphosphate
dCTP	2'-deoxycytidine 5'-triphosphate
DEPC	Diethylpyrocarbonate
DIG	Digoxigenin
dGTP	2'-deoxyguanosine 5'-triphosphate
DNA	Deoxyribonucleic acid
dTTP	2'-deoxythymidine 5'-triphosphate
dUTP	2'-deoxyuridine 5'-triphosphate
EDTA	Ethylenediaminetetraacetic acid
ELISA	Enzyme linked immunosorbent assay
EU	Emory University
F	Forward (when used in conjunction with a primer name)
FBS	Fetal bovine sereum
FITC	Fluorescein isothiocyanate
GAPDH	Glyceraldehyde phosphate dehydrogenase
GKO	Gamma-interferon knock out (mice)

h	Hours
HCl	Hydrochloric acid
HIV	Human immunodeficiency virus
hsp70	Heat shock protein (70 kDa)
ID_{50}	50% infectious dose
IFA	Immunofluorescence assay
k	Inactivation rate constant
kcal/day	Kilocalories per day
KCl	Potassium chloride
L	Liters
Log_{10}	Base 10 logarithm
Ln	Natural logarithm
LR	Logit response
LT2ESWTR	Long Term 2 Enhanced Surface Water Treatment Rule
M	Molar
mg.min/L	Milligram minutes per liter (CT units)
mg/L	Milligrams per liter
μg/mL	Micrograms per milliliter
min	Minutes
mJ/cm^2	Millijoules per square centimter
mL	Milliliters
μL	Microliters
mM	Millimolar
μm	Micrometers
μM	Micromolar
MPN	Most probable number
mRNA	Messenger ribonucleic acid
MuLV	Murine leukemia virus
MWD	Metropolitan Water District of Southern California
N	Number
NaCl	Sodium chloride
NaOH	Sodium hydroxide
N_0	Number of organisms at time zero
N_t	Number of organisms at time t
NA	Not applicable
nd	Not determined
nm	Nanometers
nM	Nanomolar
NMRI	Naval Medical Research Institute
NTU	Nephelometric turbidity units

OPR	Ongoing precision recovery
P	Probability
P	Proportional infectivity
PAC	Project advisory committee
PBS	Phosphate buffered saline
PCR	Polymerase chain reaction
pers. comm.	Personal communication
PFR	Plug flow reactor
PI	Propidium iodide
PO_4	Phosphate
QC	Quality control
r	Correlation coefficient
R	Reverse (when used in conjunction with a primer name
RFLP	Restriction fragment length polymorphism
RL	Response logit
RNA	Ribonucleic acid
RPMI	Roswell Park Memorial Institute
RT	Reverse transcription
RT-PCR	Reverse transcriptase polymerase chain reaction
SCID	Severe combined immune deficient (mice)
SDS	Sodium dodecyl sulfate
SPW	State Project Water (source water)
SSU rRNA	Small subunit ribosomal ribonucleic acid
SUVA	Specific ultraviolet absorbance
T	Temperature
T_a	Annealing temperature
TOC	Total organic carbon
Tris	Tris (hydroxymethyl)-aminomethane
U/mL	Units per milliliter
UA	University of Arizona
UDG	Uracil DNA glycosylase
U.K.	United Kingdom
USEPA	United States Environmental Protection Agency
UV	Ultraviolet

IWA Publishing's authorised EU representative for General Product Safety Regulations is Diane D'Arras, 15 rue Duret, 75116 Paris, France, e-mail: safety@iwap.co.uk.

Printed and bound by CPI Group (UK) Ltd, Croydon, CR0 4YY
17/09/2026
02218808-0015